THE

OF A JU

THE FIELD-BOOK OF A JUNGLE-WALLAH

BEING A DESCRIPTION OF SHORE, RIVER & FOREST LIFE IN SARAWAK

BY

CHARLES HOSE, Hon. Sc.D. (Cantab),

F.R.G.S., F.Z.S., Hon. Fellow, Jesus College, Cambridge; Member of the Sarawak State Advisory Council; Formerly Divisional Resident and Member of the Supreme Council of Sarawak

AUTHOR OF "*Pagan Tribes of Borneo*" IN CONJUNCTION WITH PROF. W. McDOUGALL, "*Natural Man*", "*Fifty Years of Romance and Research*"

WITH BLACK-AND-WHITE PLATES

SINGAPORE
OXFORD UNIVERSITY PRESS
OXFORD NEW YORK

Oxford University Press

Oxford New York Toronto
Petaling Jaya Singapore Hong Kong Tokyo
Delhi Bombay Calcutta Madras Karachi
Nairobi Dar es Salaam Cape Town
Melbourne Auckland

and associates in
Beirut Berlin Ibadan Nicosia

OXFORD is a trademark of Oxford University Press

First published by H. F. & G. Witherby, London, 1929
First issued as an Oxford University Press paperback 1985
Second impression 1986

ISBN 0 19 582635 3

Printed in Malaysia by Peter Chong Printers Sdn. Bhd.
Published by Oxford University Press Ptd. Ltd.,
Unit 221, Ubi Avenue 4, Singapore 1440

CONTENTS

CONTENTS

LIST OF PLATES

To

My Cousin

ALICE J. HARRISON

In Grateful Memory

and Acknowledgement

of Help and Encouragement

unfailingly given

CHAPTER I

ALL ON THE BORNEAN SHORE

A Bornean Breakfast—Turtles and Turtles' Eggs—The King Crab—A Local Turkish Bath—The Ship-worm —Strange kinds of Food—Fish attracted by Light—A Compleat Angler—Pink and Blue Prawns—Bornean Pirates—Bee-eaters—Swifts and edible Nests—A Spirit-plant—The Scaly Ant-eater—A Malay Legend —Adventure with a Bear—A Sacred Animal.

I AWOKE with the dawn and, for a moment, wondered where I was; for the early morning in the tropics has a romantic element, bringing recognition and arousing mystery. Here was I, in my own familiar corner of the world—anchored in a small sailing vessel at the mouth of a minor river on the Bornean coast; I knew very well what was going to happen, yet I was prepared for small surprises. So true is it that Nature, which is ever the same, makes all things new.

Looking through my mosquito-curtains and waiting for the coffee which my servant, with Chinese precision, was preparing, I watched the opening of day. On the shore the mist was lifting from the woods and the dew stood out clear and bold on the undergrowth. Around me I heard the cry of awakened beasts, the whistles and calls of many birds, the high treble of insects, many varying voices; beneath them all the ground-bass of the infinite Sea.

On a sandy point, a hundred yards or so away, my Chinaman is conjuring back life into last night's fire; elsewhere, the boatmen are preparing their morning rice. To supplement it, one of them has collected some *Umbut,* the unopened heart-leaves of the Nibong palm; another, more enterprising, can be seen returning along the beach after a successful search for turtle's eggs. These are a delicacy easily prepared; one boils them like any other egg, pulls off the leathery shell, and consumes them at one's ease ,with kindly thoughts of the creature that produces them, for she is perhaps the most interesting reptile of the Bornean coast. There are two kinds of turtle found here, one the Green Turtle (*Chelone mydas*); the other, the beautiful, but smaller, Hawksbill which yields the tortoise-shell and is known as *Chelone imbricata.* Green turtles are found mostly on the small islands and sandy shores of the coast; their movements are so calculated and pawky as to suggest a Scottish ancestry. Very often a prospective mother-turtle takes a long and devious road to her nesting-place, looking carefully about her before she lands. Well beyond the tide-line, she scoops out a hole in the dry sand, using her flipper like a hoe; and, having laid her eggs, returns to sea by an entirely different route. The egg-hunter, therefore, finds himself playing a game of " Hunt the Thimble," with no one to say " Hot " or " Cold."

After the hatching period, which is about five or six weeks, the babies slowly work themselves out of their sandy birth-place, and scramble down to the sea in long processions; before they reach the deeper waters where they live, many are devoured by sharks.

The turtle is a tactician, but Man, taught by experience, sets his wits against him; and in Celebes, the next-door island to Borneo, it is interesting to watch the local methods of catching them as they lie asleep on the surface of the sea, or in shallow water when grazing. A native with a rope tied round his waist dives from a boat into the water and clasps the creature in his arms; when he has got a good hold, his companions haul at the rope and bring in the two together. An even more spectacular method is one in which a small fish, *Echeneis naucrates,* having a flat sucker on the top of its head, is employed. The fish is tied tightly by the tail with a piece of fine string and dropped into a likely spot for turtle, usually in shallow water. In its efforts to escape it swims about until it lights on a turtle, to which it attaches itself by the sucker, probably in order to get a purchase. The turtle having been thus located is then collected in the ordinary manner.

Breakfast finished, we start on our journey along the coast. Malays are excellent sailors, as indeed are all peoples of Arab descent; and it is quite thrilling to watch our small ship, skilfully handled, take each incoming wave so that nothing more than a light spray comes on board. When we reach the open sea the going becomes easier and we hoist a couple of matting sails, about eight feet square, on our bamboo mast. Before we finally leave the river, however, and while the water is still fresh, we notice, among the decaying vegetation near the banks, a number of curious spouting fishes (*Toxotes jaculator*). These are a kind of perch, and get their name from their habit of stunning their prey (flies and other insects) by spouting water at them from a short distance, as they rest on an overhanging leaf. Hereabouts, especially at the mouth of a sandy river-beach, one may catch a sight of a King Crab (*Limulus*), or *Blankas* as the Malays call him. He is a strange antediluvian-looking creature, of a warm grey colour shot with green, roughly half-spherical in shape, with an extraordinarily tough shell. His tail works on a hinge and looks like a marlin-spike or a cobbler's awl; it is used for purposes of entrenchment, for this curious beast loves a comfortable dug-out in the sand, into which it settles for protection from the heat. This tail is also used to ward off enemies, and makes a formidable weapon, although there is no poison emitted into the wound. When he is properly settled in his lair, you can see nothing of him but two of his eyes. Of these organs, the King Crab is blessed with, seemingly, two pairs, anterior or median, one on each side of the raised part of the foremost spike of the carapace, which are only visible on careful inspection, and two more posterior and lateral. It seems that only one pair of these lateral eyes is used, the median having ceased to function.

The creature is said to be a native delicacy, but personally I have only tasted the mass of small eggs which, varying considerably in size, are found (of all places in this world) in what may be described as its forehead.

Dr. Gordon of the Natural History Museum acquaints me with the curious fact that the ovaries are found in the fore part of the shell or carapace of the King Crab, more correctly called the Sea Scorpion, about which very little appears to have been recorded.

In both sexes the gonads (reproductive organs) consist of systems of ramifying tubes lying in the shield above the other organs of the body. The males are smaller than the females. At the breeding season these creatures come ashore in pairs during high spring tide, the male clinging to the shield of the female. The eggs are fertilised after they have been deposited in the series of holes which the female digs for their reception. The first hole selected is near the upper limit of high tide, and the mass of eggs produced is said to be half a pint in volume.

At the mouth of the river are floating a number of great logs carried down from the back of nowhere, and riding idly on the swell. Some of these are of soft wood, and, being churned by the ocean during the north-east monsoon, turn to pulp, and form, together with various kinds of sticks and decaying vegetation, a compact mass, not unlike the leaf-mould of English gardens. For this the Bornean native has a peculiar use, employing it in a sort of Turkish bath, known as *Bertangas,* for the treatment of rheumatism. The patient is enveloped in a close-fitting mantle of this material, which covers his whole body leaving his head alone emerging from a repulsive-looking cone. Heat is now introduced from below through bamboo tubes to start a sweating process; the patient meanwhile looking very much like Mr Cyril Maude in "Lord Richard in the Pantry." As a specific this treatment is considered invaluable, and I am inclined to believe that there must be some sort of iodine salt in the decayed vegetable matter. It has a not unpleasant odour, with a tang of the sea about it reminiscent of our seaweed.

The logs of the harder woods remain more or less unravaged, except for the depredations of *Teredo navalis,* or the ship-worm, with whose burrows many are completely honey-combed. When one looks at one of these creatures (which one can do quite easily, as the logs are alive with them) one wonders how it can create such havoc. The *Teredo,* who is known in Borneo as *Temilok,* did not always live this fixed sort of life; at one time in its existence it must have been a free-swimming creature, having special organs of its own with which to guard itself against the dangers of the deep. The larva swims in all directions with extreme agility; later the external skin bursts, and, after being encrusted with calcareous salts, becomes a shell, which is at first oval, then triangular, and at last very nearly spherical. The larva of the *Teredo* possesses, moreover, sense-organs similar to those of several molluscs, and also eyes. In Eastern waters an enormous species called *Teredo gigantica* is to be found which sometimes attains the length of five to six feet and a diameter of two to three inches. This species, however, does not bore in timber or make its habitation in logs, contenting itself with boring into the hardened mud of the sea-bed.

One is rather inclined to wonder what use Providence can have designed for the *Teredo,* until one finds that most coast natives regard them as a delicacy. In the East the faculty of wonder soon dies from overwork—especially in matters of diet; lizards, sea-slugs, shark's fins, locusts and the grubs of flying ants are eaten in many places—but in Borneo and the neighbouring islands one comes across something more *outré* still. Personally, I do not see myself eating these nasty-looking worms, but they figure on the menus of the local gourmets, fried like whitebait, the head and hard shell-like jaws alone being removed. I have aften wondered, in view of the damage done by the *Teredo,* whether this custom is not part of a policy of Hate; "You eat our wood," says the Bornean, "Very well! we will eat *you.*"

About mid-day, set high on a cape, we notice a lighthouse, and I think of what one of Kipling's characters, in this very part of the world, says: "You understand, us English are always looking up marks and lighting sea-ways all the world over, never asking with your leave or by your leave, seeing that the sea concerns us more than anyone else." But there are other people concerned with lights besides ourselves; and the fascination which a bright light has for fish, is turned to advantage by the coast-people, when searching for crayfish in these shallow transparent waters. They stalk the creatures with a sort of barbed spear about seven feet long, which they push in front of them along the bottom. In order to make manipulation easier and keep the spear more or less stable, a sort of strut is attached at an angle and reaching down to the sea-bottom; while about two feet up the shaft and just above water-level, is fixed a candle which serves a double purpose, attracting the crayfish and guiding the fishermen.

Fishermen other than human seem to have discovered this use of a light, for there are certain creatures known as "Angler-Fish" which use this method. Of these fish, for a description of which I am indebted to Dr. Tate Regan, F.R.S., of the British Museum, there are about fifty different kinds, exhibiting many varieties of form and size, but uniform in this, that they all have the first ray of the dorsal fin placed on the top of the head and modified into a line and bait. One group of these, known as *Ceratioids,* live in mid-ocean, about half-way between the surface and the bottom. In these gloomy waters their bait is a luminous bulb, either set directly on the head, or else at the end of a long

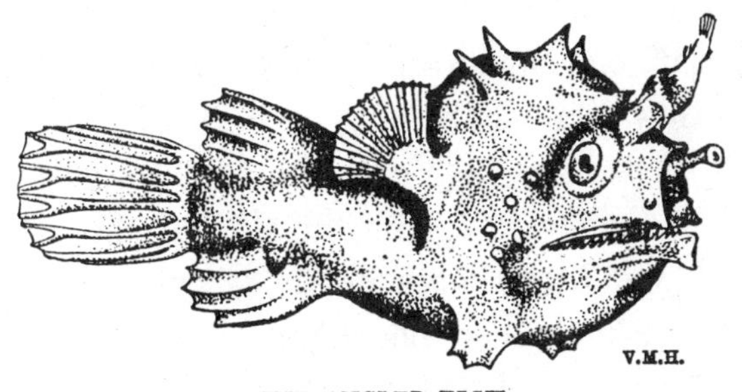

THE ANGLER-FISH

filament, in some cases projecting from the skull and looking like a rod and line. In one species (*Lasiognathus*) the line actually extends beyond the bait, and ends in a triangle of hooks. Here indeed is the Complete Angler.

Perhaps, however, the most remarkable peculiarity of these Oceanic Anglers is that all the free swimming-fish are females, and that all the males are dwarfed and parasitic on the females. The habits and conditions of life of these fishes, solitary, sluggish, floating about in the darkness of the middle depths of the ocean, make it evident that it might be difficult for a mature fish to find a mate.

This difficulty appears to have been overcome by the males, as soon as they are hatched, when they are relatively numerous, seeking the females, and when they find one, holding on to her for life.

The males first hold on by the mouths, then the lips and tongue fuse with the skin of the female, and the two fishes subsequently become completely united; the male being nourished by the blood of the female and the blood system of the two being continuous.

The Ceratioids are unique among back-boned animals in having dwarfed males, and in having the males nourished in this manner.

As we pass along, clumps of Casuarinas and Screw-pines stand out boldly; while down the bare cliffs tumble glistening amber-coloured waterfalls. I always love stopping at one of these falls, for the water gives a thrill like nothing else in the world and both soothes and invigorates; after it one can the better enjoy the rich colours of the seascape, and (to be materialistic) the pleasures of lunch. As we watch our make-shift sail bringing us nearer to the chosen spot, our steersman (I always make a point of having the same man, when possible) points to a shoal of porpoises, or perhaps a shark, following a few yards in our wake; the shark is probably as ready for his food as we are. Overhead the white Sea-Eagles and Brahmany Kites whirl in great circles, every now and then darting down at the sight of a large fish, each vying with each in the grace of their sweeping motions. On the shore the locals (optimists, surely) shout and wave their hands at them, trying to make them drop their prey.

Lunch over, one walks leisurely along the shore, where, if shrimping is in progress, the curious sight may be seen of millions of tiny shrimps passing along in line through the shallow water. As the rays of the hot sun shine down through the water, the whole seems to have changed to a beautiful pink. When I first saw these numerous tiny pink prawns, I was reminded of the artist who painted the live lobster red and wondered if I myself was falling into the same delusion. However, on examining some, I found that the bodies of the prawns were of the ordinary greyish-brown colour. Their legs, however, were red and there are so many of them that, perhaps together with refraction, they tinge the whole mass of water a delicate pink.

At certain seasons, the natives catch them in close-meshed nets fixed on a pole, much like our shrimping nets, which they push to and fro in the shallow water along the beach. Basketfuls of them are spread out on mats and dried in the sun. They are eaten with rice, or else pounded up with salt into shrimp-paste; for a month or so in every year the whole atmosphere reeks with the smell of shrimp-paste while this favourite Malay article of diet is being prepared.

A similar Malayan delicacy, but one which appeals to Europeans, is a larger kind of prawn (*Palæmon*), found in the fresh or brackish water at the mouths of most of the rivers. This creature is almost as large as a lobster and has two very long blue claws. It is caught either in box-like traps made of bark, or else in cast-nets—in either case a ground-bait is used consisting of grains of rice mixed with some tasty and strong-smelling preparation. A piece of stinking decayed coconut flesh (copra) placed in one of these lobster traps proves a most attractive bait.

Along the rocky foreshore one often sees beautiful yellow ground orchids of the Calanthe family, growing on the sides of the cliffs and apparently without much soil. They have a beautiful spike of golden flower rather like a spiræa, and large flat leaves.

As one wanders along these wild, rocky coasts with their beautiful caves and luxuriant plant life it seems almost impossible to associate their majestic beauty with scenes of a time when pirates ravaged the seas, and killed and warred against the peoples who inhabited the shore, or captured and sold them into slavery. Yet it was in the creeks and rivers of this northern coast that the notorious Illanun pirates of the eighteenth century " dug themselves in," a pest so formidable that the British Admiralty considered it " certain death " to venture into their strongholds; and it was here and hereabouts that the famous Rajah Brooke and the "Little Admiral," Sir Henry Keppel, earned the thanks of civilisation, smoking them out as one might a nest of hornets.

Pondering on this past history I sat down to think. But my thoughts were disturbed, for I noticed that two beautiful birds (Bee-eaters, or *Merops*), which had been flying to and fro about me, had alighted on the ground near by. They are migrants and are known to the people of Borneo as *Burong taun,* which means "the Annual bird." Their colouring is, in the main, green; but the neck, the upper back, and the wings have a broad colouring of ruddy brown; the lower part of the back is buff, and the long middle feathers of the tail are tipped with black. The forehead is pale green and white, and the throat bright yellow.

I watched the pretty things for some time, and then turning my eyes for a moment to a vessel out at sea, found they had disappeared. I had been looking in the same direction all the time and should have noticed had they flown away, so I got up and walked slowly to the spot where I had first seen them, and, to my astonishment, discovered several small holes in the ground, from one of which, while I watched, a bird flew out. When one considers the beak and the feet of these birds, one is not inclined to think of them as capable of making holes; however, picking up a piece of stick I pushed it down the hole to see how far it would go, suspecting that these birds were nesting. The hole ran parallel to the surface at a depth of a few inches for a distance of a foot or so, and at the end I discovered a nest with five glossy white eggs. These eggs were almost exactly like those of a Kingfisher. A little later I noticed that the Casuarina trees all about seemed to be covered with these birds, apparently mating.

In the forest regions of the Hinterland there is another species of Bee-eater, a fairly large bird, about the size of a blackbird. Its plumage is green, with a pinkish-red head. I have never been able to find where the forest variety nests, but as I watched these little coast birds of the same genus I could not but think it likely that, since the sea-coast Bee-eater nests in a hole in the ground, like the Kingfisher, his brother of the forest must build his nest in the hollow of a tree or a rotten log, or, possibly, in the ground, where the roots of a fallen tree may have exposed the soil.

Continuing our journey slowly along the coast, we observe a number of small caves of sandstone formation, similar to the limestone caves found farther inland only rather smaller. Here a wealth of vegetation seems to spring from nowhere but solid rock. Along the shore, adhering to the walls of the caves, may be seen the nests of the swift known as *Collocalia linchii.* This is a small bird, mostly black in colour, with a white-speckled chest, which builds a nest of moss and straw, smeared over with some gelatinous substance by which the nest is made to adhere to the rock. Owing to the nest not being entirely of this substance it has no important commercial value to the natives, who collect other edible swifts' nests for the Chinese market. This little swift is considered rather a nuisance by the natives; for he is a tyrant and has a habit of frequenting the haunts of the two other species of swifts, the *Collocalia lowii* and *Collocalia fuciphaga,* and disturbing them, occupying the nests and more suitable positions and ousting the more valuable species.

Along marshy plains which sometimes occur on the coast, and which are undoubtedly long-forgotten clearings of the forest, short sedge grasses may be seen growing. Here one finds the Drosera, a carnivorous plant not unlike the two European species found in the fenlands of East Anglia. The little plant is about the size of a two shilling piece and has the habit of catching flies and insects in its leaves. Another fairly common plant with similar carnivorous habits is a species of Pitcher-plant (*Nepenthes*), more or less a dwarf variety of the large beautifully-coloured jungle species. What was of greater interest was a species of Myrmecodia, a strange plant which seems to have struck the natives as something weird, as they call it *Anak hantu,* or " child of the spirits." At the basal portion of the stem it has a sort of hump, capable of storing dew and other moisture. The skin of this growth is so thick that it prevents evaporation, and, if the stems are cut across, they are found to be full of tiny catacombs and channels in which ants constantly shelter, breeding in prodigious numbers and making their permanent home there—for which reason the plants are called " hospitating plants "—and it would seem that the biological connection which exists between them and the hospitating ants is a case of symbiosis for reciprocal advantage, the plant apparently entertaining the ants because they are useful as a defence against other insects and various hurtful creatures.

Quite a number of good-sized trees are found about here, growing right down to the beach; among them the *Hibiscus tiliaceus,* not like the " shoe-flower " that one sees in the Scillies and other sub-tropical climes, but a great tree as tall as a house, and with a trunk eighteen inches or more in diameter. It is an interesting tree, for from its inner bark the natives obtain a useful string known as *baro,* that procured from the younger branches being superior as the bark is more tender and pliant. Although in considerable use locally, it has never yet been turned to commercial use to any extent; but probably, in judicious hands, it might prove a valuable commodity. The large flower of this Hibiscus is of a beautiful yellow, which turns to a ruddy tinge as the tree grows older.

Before turning away, for we had to move on towards our evening's camp, I came upon a specimen of that extraordinary creature, the Ant-eater, or Pangolin (*Manis*), in the act of devouring a nest of red ants, on the bough of a tree. Having lost its teeth through long ages of disuse, this creature has a very long tongue with enormous salivary glands, the sticky secretions of which enable it to pick up the ants. Its chief means of defence are its sharp strong claws, which are used to tear an ant-hill to pieces. Its back is covered with a mass of hard horn-like scales. It is, however, able to emit at will an extremely unpleasant odour, enough to baffle the most hardy pursuers.

My friend, the late Robert Shelford, in his book, "A Naturalist in Borneo," repeats an amusing Malay legend (which may or may not be true) about the Manis. When there is a shortage of his natural prey, he lies down in the forest, curls himself up into a ball, and shams dead. Soon thousands of ants, acting perhaps on a sort of "bazaar rumour" of the jungle, flock together to feast on the supposed corpse, and swarm under the slightly raised edges of the Ant-eater's scales, in order to get to the soft skin underneath. "When the Manis considers that it has collected sufficient numbers of the ants, the corpse comes to life again, straightens itself out, and in doing so shuts down the scales and imprisons the ants. It then trots off to the nearest pool of water or stream, into which it plunges and arches its back, thus raising the scales again. The ants float off on to the surface of the water and are licked up with the long slender tongue."

I once captured one of these creatures and put him in a large wooden box, which I nailed down until I could find him a more permanent residence. My night's rest was broken by the noise made by him trying to escape; and next morning I found that he had used his claws to such effect that he had torn away about half the lid, and had escaped.

The Ant-eater is a slow mover on the surface, but a quick burrower; it makes its home in holes in the ground or in hollow trees. In Borneo we have the long-tailed species only (*Manis javanensis*). He is probably a remnant from a prehistoric age, and, I am bound to say, he looks the part.

On one of these coastal voyages, before making a start in the morning, I turned aside with a couple of my Iban followers, to examine a tall *Tapang* tree, the larger branches of which were literally covered with clusters of wild bees clinging to the underside, at least a hundred families. When I reached the foot of the tree, to my astonishmnt a half-grown honey-bear—about the size of a large bull-dog—came sliding down the trunk, tail foremost, growling fiercely, and using his powerful claws as a brake. The last few feet he accomplished by rolling; but he rose rapidly, growing angrier every moment, and then reared up on his hind legs as if he wanted to fight me. He reminded me rather of the way Mr. Snodgrass "announced in a loud voice that he was going to begin"; but he evidently thought the better of it, for he sheered off to behind the spreading roots of the tree, and then with a gesture of contempt and a final snarl, ambled off into the forest.

When I looked round, I found that my Ibans had disappeared. I knew that they alone of the Bornean tribes hold bears in respect, but I asked them why they had fled. They explained that it was strictly *mali* (taboo) to injure a bear, or even to attempt to do so, as for instance by laughing at one. I asked them if they would eat bear's flesh, but they seemed horrified at the very idea. They look on the Bear as one of their omen-animals, and the Bear to most Ibans is as near an approach to Totemism as anything that I have ever found in Borneo; and indeed in their attitude towards the animal one finds Totemism in the germ. The Bear is, for some individual Ibans, a Gnarong, a sort of guardian angel, specially belonging to one person much like the Roman *Genius,* or the Egyptian God " who walks by the side," or even the δαίμων of Socrates; a personal helper and protector taking the form of a chance-seen animal. Now some Ibans have a tribal protecting spirit named Impernit, who, apparently, was an actual chief of the Undup river; for I have seen his tomb (*Rarong,* or *Sungkup*)—built above ground on piles,—in which there is a tube inserted as a means of communication between this world, and the World after Death; here the chief's body lies, as the Ibans say, " to be with us even in death." Impernit was probably a great warrior some generations back; and the probability is that the Bear was the Gnarong of this tribal god, and is for that reason held sacred by the whole tribe.

CHAPTER II

A 'LONGSHORE VILLAGE

"Mudskippers"—A Fish that was Drowned—A Coastal Village—Crocodile Turned Burglar—Some Strange Crabs—"The Porter"—A Miscellaneous Catch—Sea-horses—Sting-rays—The Balloon-fish—Catfish and Barbels—Megapodes and Their Mound-dwellings—Treasure from a Wreck—Boom, an Important Chief—Rajah Gamiling—A Sea Festival—The Story of an Image.

TOWARDS evening a headland comes in sight, with a tiny village hidden among groves of cocoanut palms and fruit trees, at the mouth of a little stream. As we near the shallow, sandy entrance of the little river, a shoal of grey mullet (*Mugil*) spring out of the water, for we are disturbing them on their feeding ground. Occasionally, I have known one or more of these fish to jump into the boat in their excitement and eagerness to escape. As we paddle into the mouth of the stream, one of the crew in the bow seizes a cast-net and throws it with great skill over them. Having thus expeditiously obtained our supper, we anchor near the bank, and moor our boat by a rope or length of rattan to a handy tree or log, and, as soon as the crew have lifted out all the necessaries, lie down for a peaceful rest, lulled by the gentle murmur of the waves or the curious droning sound of a prawn as it grates against the bottom of the boat, while the shrill notes of little shore plover, which run along the beach, and of other small sea-birds, ring in one's ears. These shore plover are pretty little birds with largish heads, which run along a few steps very quickly and and then stop. They are usually of a greyish-black plumage on the head and upper parts of the body, with pure white breast and under parts. Their note is shrill and clear, and they congregate in large numbers on mud-flats at the mouths of rivers. As one sits quietly on a log by the muddy shore, strange sights occur, most notably certain curious little amphibian fish, which flop down the banks into the water in a series of short jumps, using their tails as a kick-off. These are Bommi fish (*Periophthalmus*), also known as "Mud-skippers" and "Jumping Johnnies." Their bodies are like those of ordinary fish, except that they can use their fins and tail for climbing and jumping; but their heads have, set high on the top, a pair of great protruding eyes by which they can perceive the approach of danger from any direction, and from which they have acquired their name.

The habits of these fish (which are found throughout the tropical regions of the old world, from the West of Africa to New Guinea) are even more curious than their appearance; for they are fast becoming land-animals, and seem only to resort to the water for breeding purposes, although they never go far away from it. When they are full-grown, they like particularly to lie in the mud with their tails in the water, and it is thought that in some way they are able to oxygenate their blood through the vascular tail-fin, which takes the place of the gills of ordinary fish. This hypothesis would, however, hardly explain the ability of the fish to remain for long periods at a time quite a respectable distance away from the water, on a tree-trunk for instance, or a mangrove log. A story is told about myself that I once tamed one of these Mud-skippers, who used to follow me about like a dog, until one fine day it fell into the water and was drowned.

I am not prepared to vouch for this story, although in my time I kept a number of strange pets; for the creatures are very shy, and very hard to approach at all, on account of their curious eyes; Mr Shelford notes that at the least disturbance they "rush towards the sea, the flapping of their bodies and tails against the wet sand making a noise like the squattering of ducks in mud."

Across the water a pleasant tinkle of gongs announces the presence of a typical coast-village. These are commonly built on poles of bamboo or the stems of palms, driven into the mud at a point just out of reach of the highest tide. The poles are lashed together with strips of rattan threaded in and out, and the floor consists of strips of split palm stems on cross-poles, kept in place by the same means, and covered with large palm-leaf mats, light but very durable. Under the floor and between it and the ground or water (as the case may be) hang crates made of rattan for the housing at night-time of the many domestic fowls which every family keeps.

Thus secured, the birds are out of the way of the household, but very much in the way of the local crocodile, as I can personally testify. I was once camping in a little village when one of the chief men of the place came to me in some concern, and with a story of damage caused by one of these brutes. If appeared that, attracted possibly by the chickens, a crocodile had swum up during the night and had broken through the floor of the house. Hearing a noise, the owner of the house said that he struck a light and saw the huge beast sticking half-way through the floor and with its immense jaws open. He could hardly believe his eyes, but hastily waking up his wife and children, fled with them into the next room, which was partitioned off. Snatching a spear from the wall, he rushed in at the crocodile, stabbed it in the chest, and fled, leaving the spear buried in the reptile's body. Whereupon the crocodile retired, leaving a gaping hole in the floor. As the story sounded true, though improbable, I went down with my informant to view the field of operations. There was a large irregular hole in the palm strips of the floor, two or three feet across, and I could quite easily understand what had happened, although I had never before heard of a crocodile entering a human habitation. The curious part about the incident was the distance between the water and the floor, which was about five feet; and how the creature reared itself up and got a foothold I cannot imagine. I came to the conclusion at the time that the crocodile was not after any human prey, but in search of chickens, for there were some crates of birds inside the house, but perhaps the wish was father to the thought, for I was myself in the habit of using similar houses, and also those built on rafts.

Blue-limbed crabs of beautiful hues lie about the mud banks of the mangrove swamps, while clumps of graceful Nipah palms, intermingled with the mangroves, are the homes of little bitterns, night-herons, and egrets. High above, the sounds of life all around can be heard, the cackle of white sea-eagles and the squawk of the laughing kingfisher as he flies along the river bank. He is a beautiful bird, with back of a gorgeous blue, buff breast and long red beak, and as he darts to and fro his peculiar call is as startling as it is distinctive.

On the stems of the Nipah palms, below tide mark, oysters of a fair size may be found, but they are not so good as those obtained from among the rocks. Then there is a kind of crab, not unlike a large scorpion (*Thalassinus*), which makes little mud hills at the mouth of a tunnel something like our mole hills; it lives only in the soft wet mud, where the tide ebbs and flows. During the heat of the day it lives at the far end of the hole, which is perhaps four feet or more in length, and passes the day, apparently, without searching for food. I endeavoured to catch some of these creatures by inserting a long thin twig into their holes. The crabs angered at being disturbed snapped at the end of the twig with their claws, and could be drawn out and captured, their obstinacy proving their ruin.

A more interesting, and even a ridiculous sight, is that of a small crab known as *Dorippe facchino,* or the Porter. It is so called because in the claws of one pair of legs it holds an oval gelatinous plate, on which there grows a little Sea-Anemone; this the crab carries about above its head like an umbrella, or a parcel from the luggage-van. It seems to be a case of mutual help; for the Sea-Anemone has stinging powers which would render it a nasty mouthful for fish or other enemies of the crab. What the Sea-Anemone gets out of it, I cannot say; nor is it easy to imagine how the partnership can have started. How does the Crab know that the Sea-Anemone can be useful to it? and how does he first pick up a young unattached Sea-Anemone?

Glancing around from our camping ground, one sees a number of long palm-poles sticking up out of the water. These are the remains of houses which have long since disappeared. If one taps one of these poles two or three small bats are sure to fly out, for the pith centre has long ago been destroyed and the hollow pole provides the creatures with a home where they can remain in peace and darkness during the daytime.

The sea now being perfectly calm, native fishermen can be seen farther along the beach engaged in catching fish with a drag-net. On approaching them we see that one end of the net is carried out to sea by some of the party, who wade in to their necks, and are just able to hold on to the net while those in a boat pay it out in a half circle until the entire net has run out. Then, moving round, they complete the whole circle, and, closing in at either end, the two parties with their united effort drag the net ashore.

Anxious to see what sort of catch they have made, we move closer, and in the pocket in the centre of the net we find many beautiful edible fish, such as bass, horse-mackerel (*Trachurus*), walking-fish (*Antennarius*), mingled with sting-ray (*Trygon*), small sharks, and various kinds of cat-fish. Among them were one or two horrid, poisonous, spiny little fish such as the sea dragon (*Pegasus draco*), the horned trunk-fish or *Ostracion cornutus,* and *Tetrodon* or Balloon-fish, and a small fish known as *Ikan iblis* or devil-fish, a curious creature with many sharp spikes which often cause great pain to the natives if their bare feet come into contact with them in the water. The Pipe-fish, also called needle-fish (*Syngnathus*), is common along the shores, and may be included in the catch. The flesh of certain species of Plotosus (the Semilang of the Malays) has a good flavour, but it is rather indigestible; a very fine glue is made from its swimming blades which looks like a sort of isinglass. This fish is of a greenish colour, and much eaten by other fish; sometimes it reaches a considerable size—specimens seven feet long and weighing as much as sixty pounds have been captured. The horned trunk-fish (*Ostracion cornutus*) is a curious-looking creature which is constantly brought up in the fishing nets. They are called Pectognathi because their jaws are coalescent. The family of Trunk-fishes are known by the curious structure of their external surface, composed of a series of hard scales forming a continuous bony armour. The body is either three or four-sided, and covered with this solid coat of mail in which the scales or plates are six-sided, and this armour is pierced with holes through which protrude the mouth, tail, and fins. The entire interior structure is modified in accordance with this external, inflexible cuirass. If we compare the general form of this creature with that of certain reptiles, the analogy between the Trunk-fish and the Tortoise is too close to escape observation. None of these fish are in demand as articles of food, and, generally speaking, are supposed to have a poisonous effect.

Here also we find the *Hippocampus,* the little " sea-horse " or " sea-dragon " common in many European and tropical seas, and sometimes found on the British coast. These fishes have but one dorsal fine, set far back, and capable of being moved in a marvellously swift fashion which reminds one of a screw-propeller, and evidently answers a similar purpose. The tail of the sea-horse, stiff as it appears to be in dried specimens, is, during the life of the creature, almost as flexible as an elephant's trunk, and is employed as a prehensile organ to anchor the sea-horse to weeds or other fixed objects. Some of these sea-horses can be seen in the new Aquarium at the Zoological Gardens in Regent's Park, Sir Peter Chalmers Mitchell having brought the original supply from the Mediterranean by aeroplane.

I next examined some of the large sting-rays (*Trygonidae*). Of these there are several kinds, one of which will ascend the rivers and fresh-water streams far beyond where the tide makes, sometimes as far as a hundred miles; the other species is a dangerous brute, for the poison causes terrible pain and a swelling of the limbs. I have known cases where natives have been unable to use their legs for weeks. The actual sting is a long bone-like spike, set on the whip-like tail and running parallel to it, usually about a foot in length, but sometimes as long as a walking-stick. It gradually tapers down to the point which is of a finger's thickness, and is divided along its length by a groove or channel, down which the fish emits its poison. Unlike other Rays, Skate and Thorn-backs, the young are born alive, not hatched from an egg.

Perhaps the most curious of this collection is the *Buntal* or Balloon-fish, which, as soon as it is landed, or comes into shallow water, begins sucking in air with strange noises until its skin is completely blown out. This *Tetrodon* or four-toothed fish, normally some six inches long, but about the size of a goose's egg when inflated, and resembling a gigantic horse-chestnut, has quite an array of soft spiny points on its skin which project in every direction when it is filled with air and blown out. Both jaws are divided in the middle, giving the fish the appearance of possessing four sets of teeth, two above and two below—hence. its name *Tetrodon.* When the balloon-fish is inflated, it turns on its back, and, if in the shallows, floats belly uppermost. Not until the air has once more escaped can it immerse itself below the surface. There is one species beautifully marked with golden stripes. The balloon-fish is not considered fit for food by the natives of Borneo, and, for this reason, when it is caught in the nets—inflated to its fullest extent and being quite useless—they stamp violently upon it in disgust, causing a devastating report like a boy bursting a paper bag.

The last of the catch to be inspected was a Barbel (*Arius*). This is a funny-looking object, possessing numerous tentacular appendages which adorn its mouth like the exaggerated whiskers of a cat, and one of which is especially long—though I never discovered whether the fish takes its name from this fancied resemblance to a cat's face, or from its manner of protecting its young, which take refuge in the mouth of the mother. These catfish lay their eggs in a secluded spot, and, should any other fish prey upon the young, these seek safety in the mouth of the adult, and are constantly so found at an age when the belly of the young fish still shows a remnant of the yolk adhering to the under part. If the catfish be held up by the tail, half a dozen young fish will drop out from the mouth of the mother. I thought this so peculiar that I wrote a description to Dr. Günther of the British Museum, and sent him some fish preserved in spirit with the young still in the mouth. In reply he wrote to me saying that he was very pleased to have my letter since it confirmed a fact which he had himself described with regard to these fish as far back as 1846.

It was on an island off the coast that we came upon that shy bird, something like a small turkey, known as a *Megopodius cumingii.* It is not often found on the mainland, but in the forest near the shores of uninhabited islands, and is caught by the natives by means of a trap, consisting of a spring formed by bending over a sapling with a noose at the end. When hunting for ground birds, the natives build a low hedge of undergrowth, and at stated intervals they make " eyes " or openings. The birds try to negotiate this little obstacle without taking wing, and run down along the side of the barrier till they come to an opening. Over each " eye " a sapling is bent, and the noose laid round and over a small platform arranged in the opening. The Megapode, taking the line of least resistance, steps on to this platform, which gives way, thus setting free the noose and allowing the sapling to spring back—with the quarry caught by the legs in the noose, by which it is swung into mid air as the sapling straightens. This bird, dull brown with huge claws, has a very melancholy cry, uttered in a single note in monotonous succession, and is called in Malay *Menimbun,* " Mound-builder."

The birds usually come out in the morning or evening and build their nests, beyond tide mark, in the form of a mound in which they deposit their eggs. Having done this, the mother troubles no further, for the young hatch out by the warmth of the sun and make their start in life alone. In digging for Megapode eggs one usually finds in the mound rotten logs of drift-wood washed up during the North-east monsoon. These are the foundation of the mound or heap, which is then built up of sand mixed with leaves, branches and twigs. The mound is sheltered, and indeed almost hidden, for the bird has many enemies beside his chief one, Man. The nesting season is apparently from September to November. The mounds are usually from six to eight feet in diameter, and two or three feet high, with burrows or runs constructed inside them. Wild pigs forage for these eggs, and often the young, on emerging from the burrow, falls a victim to them or to large lizards of the Varanus variety.

The runs are often excavated along the sheltering side of one of the foundation logs, and I believe that these serve as entrance galleries for the hen, who goes inside the mound to lay one egg a day. I believe, also, that many of these mounds are communal, and shared with other families; which, if it is the case, is curious, because of the evident analogy in the communal Long Houses of the natives tribes of Borneo.

When the birds have finished laying, they carefully hide their tracks by scratching up leaves and sand and other debris so as to close the entrance to the mound.

Such are some of the curiosities which face the voyager on these shores of the big island, but interest is not confined to what is vulgarly called the Brute Creation. Romance is by no means lacking, for here, between Baram Point and Bintulu, some years ago, evident traces of a wreck and indications of buried treasure were found. The Government of Sarawak (in Sir Charles Brooke's time) took up the matter, but finding that Jesuit missionaries had been on the ground before them, in the early part of the eighteenth century, and in 1896, with no visible results, they quitted what was evidently a barren field.

However, if there are no secrets from the past, the present is amusing enough; for among these little villages live a tribe of expert boat-builders, whose small clinker-built schooners, thirty to fifty feet in length and two-masted, ply a thriving trade between the Chinese and Malay towns of the littoral. I made great friends with one of these, a Milano named Boom, a name apparently derived from the discharge of a gun, for he was a chief of some importance, and may have been accorded a local salute.

Boom was not merely a boat-builder, but also an expert carver, having gained a considerable local reputation for the making of images used in festivals in honour of the God of the Sea (known as Raja Gamiling), in which, more by token, he officiated as Expert, Prophet or Thaumaturgist.

The Gamiling festival (known as *Gnima*) is a sort of mixture of Science and Superstition, motives not unknown to our own Pilchard and Mackerel-fishers. At the season when fish are expected, an image of Raja Gamiling is set up on the foreshore with offerings of fish and fowls. Then little rafts of Nipah stems, called *jong,* are built and let drift out to sea, bearing streamers of Areca-nut blossom, coloured rice and eggs. The idea of this ceremony is the propitiation of the Spirits; but there is more in it than that, for Boom used to take up his position on the shore and watch carefully which way the rafts floated and which way any current might be setting. The course of the jongs, which were despatched daily for some ten days, was carefully marked by a sort of channel of leaf topped casuarina stakes planted in the shallow water; along these in the course of time, the fish would be guided along the swift in-coming tide into fish-traps (*pingiris*) set for them. It is therefore easy to see that Boom, as Tribal Chief, Village Artist, and Magician, was a person of some importance to the fisher-folk.

Once, when I was poking about that part of the world, I came upon one of these images, derelict, and I recognised it as Boom's work; it had a sort of half shapeliness, reminding me in a vague way of Mr Epstein's work. I showed it to my friend, and he acknowledged it as possibly his handiwork, but refused to take it back; it had served its purpose, said he, and had never been intended as a thing of beauty and a joy for ever. I was not sure how far he was serious, and when I was next on leave in England, I had a cast made of Gamiling in copper, and offered it to Boom when I next met him. To my surprise he quite firmly, but quite politely, refused my present; he again took the line that its work was done, and added that, in any case, the cast had not the same virtue as the original, which had a sacerdotal value.

I suppose he was right. His scruple deserved respect, neither less nor more than the punctilio which all right men feel concerning the instruments and symbols of their faith, and if I did not fully understand his feeling, to question it seemed an impertinence.

CHAPTER III

"OLD MAN RIVER"

Ille terrarum . . . The Great River—Sharks—Fish-traps—The Water-buffalo—The Egret—John Chinaman—The "Adeh" and her Officers—The Hidalgo of the Seas—Friendly Bees—Gadflies—"The King of the Jungle"—Some Interesting Pigs—Baram Residency—A Tropical Paradise—Lizards—Mason Wasps and Carpenter Bees—Fire-flies—The Professor and the Mosquitoes.

FOR me Borneo means the Baram River and the Baram Division. It was there that I was first posted as a cadet, as long ago as 1884; and, for an added claim, the whole region is one of the very richest in all natural beauties, and when I first came upon it, was almost unknown territory. Small wonder, therefore, that I should keep a soft place in my heart for this corner of the earth.

Not so little a corner either; for the Division is only just smaller than the whole area of Wales, and the distance from the sea to the head-waters is all of three hundred miles. Even at the mouth it has the appearance of a notable stream, for the volume of water carried down by its extensive and intricate river-system makes for grandeur; at no season will you find navigation easy across the sand-bar at its mouth, and during the North-east monsoon its waters are closed to vessels of any but the shallowest draught.

The lower reaches of the main stream, for the first fifty miles or so, are lacking in distinction, but at the mouth itself there is plenty to entertain the traveller and the naturalist. Here is the home of several species of small shark, although they are occasionally visited by individuals of the larger sizes, such as are found about the islands north of Australia. Concerning these monsters, Mr. J. Hamley, well known as a shark-fisher, tells me that one of the great man-eating sharks, known as the Tiger Shark, was seventeen feet in length and weighed 1658 lbs. The liver weighed 397 lbs. and the oil produced from it was 342 lbs. in weight, or 38 gallons. The salted flesh realised from £35 to £40 per ton and the fins from 3/2 to 10/-, these being about 1½% of the gross weight. Compared with such monsters, the Baram sharks are small fry, but they are big enough to be unpleasant. About them there is a local saying that anyone who makes a tinkling noise with a pot or a pan will attract the brutes; the Borneo Malays got so far as to consider it necessary to keep absolutely quiet when in a shark-infested area. There is probably a good deal of truth in the tradition; I myself on one occasion (when perhaps there was a certain amount of noise going on) saw a hammer-headed shark about five feet long suddenly rear itself out of the water quite near the boat. Curiously the fish, owing perhaps to some effect of light, appeared a bright red in colour, which did not add to its attractiveness.

Sharks do not seem to cause much trouble to the smaller fish in these waters, for there are plenty of various kinds, some of which are very good eating. The natives about here catch most of their fish in traps known as *Kilong*. A Kilong is really a sort of spiral maze, perhaps ten yards in diameter and consisting of a fence of close-set poles driven into the mud or sand to a depth of a foot or so. The fish enter the maze on the ebb tide and work their way round to the centre of the spiral; then, when a sufficient number are collected, the fisherman lets down a sort of sliding shutter and imprisons them, taking them out at his leisure in a hand-net. Above these traps are always seen a number of birds hovering high in the air, eagles, hawks, and kites, and occasionally the Frigate or Man-of-War Bird, a creature with a slender body of a dark-brown colour, a short neck and strongly hooked beak; it is said to portend a stormy sea and bad weather.

On the shore and towards Baram Point, one often comes across various kinds of migratory birds, doubtless attracted by the lighthouse, which is built on land cleared of forest and drained. The light is a powerful one and can be seen eighteen miles out at sea, and during the North-east monsoon attracts birds flying south from the continental cold; no doubt they find it a very convenient resting place, for the nearest land northwards is China, more than 1000 miles away. Another favourite haunt of these immigrants (starlings, pratencoles, and so forth) is in the coconut and casuarina trees, of which there are a great number here, and which at the right season are as noisy as an Eastern bazaar on market day.

In the swamps and little backwaters (known locally as *trusans* or *bruans*) one always sees a number of wild water-buffaloes, smaller than the Indian species, but with fine horns. They are particularly fond of this locality because here even in the driest season, when everywhere else is more or less parched, they can find a nice comfortable bed of slush to wallow in. Here they lie, enjoying a luxurious ease, with their faces and horns alone showing above the surface of the water, so that at first glance one hardly sees them at all. One might think that these fat, stupid creatures would be an easy target for anyone who could hit a haystack; but the fact is quite the other way. They possess an uncanny sense of smell, and, directly they scent danger, they shamble off into the underwoods, showing a surprising turn of speed; approaching them down wind you will never get within a quarter of a mile of them.

Curiously, this shy animal proves, when domesticated, most docile, a small boy being able to lead it where he will by means of a piece of pack-thread passed through its nose. Apart from its food value (for its milk is said to be thicker and richer than cow's milk) it makes a very serviceable, though slow, draught animal, and is also used for agricultural purposes. The plough is practically unknown in Borneo except among the Dusuns of the extreme north, who may have learned its use from the Chinese, being themselves of partial Chinese descent; but in the rest of the island its place is taken by the buffalo. The process is a simple one. The mud-flat chosen for the crop (rice) is first cleared of grass, scrub, and other vegetable matter, which is then stamped with the slush. The owner then takes three buffaloes and leads them by rattan cords passed through their noses, up and down the field, which they churn up with their ungainly feet. When the land has been thoroughly stirred up in this way, the water is drained off into a field at a lower level, where the same process goes on. The rice is then planted in the *purée* thus obtained before it has time to settle.

In British North Borneo the buffalo is put to a use which sounds comical, but is really very practical; he is used as a mount for the district police, who find these unwieldy animals reliable and cheap; and, after all, appearances are not everything.

The Bornean buffalo has only comparatively recently been domesticated and trained; until about eighty years ago the larger Indian buffalo was used. Now, however, one finds both kinds and also crosses between the two species; while occasionally an albino buffalo, with a pink skin, may be seen.

On the sandy beach above the level of the tide are found numbers of plants of dwarf habit belonging to the vetch family, with a reddish-purple flower and a strong wiry stem as thick as a lead pencil. These are used by the natives, when rattan is not available, for twine and ropes. The banks of the river within the bar are crowded with a profusion of Casuarina, Mangroves, Nipah, and, towering over them all, the tall Nibong palms—all commercially valuable. Among them and under the overhanging fronds flits that beautiful night-heron, the egret, also known as the paddy-bird. The fully grown bird is uniformly white, but the young are covered with a brownish-yellow down, not unlike the colour of a baby Aylesbury duck. Whole flocks of these young birds may be seen during the North-east monsoon, looking as if they came from the north, for, so far as I know, the egret does not nest in Borneo itself. In Sarawak the birds are, very wisely, protected, for they do a great deal of good in destroying certain parasites found on cattle, especially the buffalo. It is indeed quite a common sight to see one of these birds perched on a buffalo's back and apparently on perfectly friendly terms with it, making a meal of the ticks; the buffalo, meanwhile, standing motionless, apparently in stolid contentment at being rid of its unpleasant lodgers.

When the Baram River was first cleared of pirates and opened for trade (which was not many years before I went there), Chinese merchants in their high-pooped junks found it worth their while to make the long journey south from the Canton River, laden with salt and with deck cargoes of pigs, fowls, ducks and other live stock, seeking a favourable market. For weeks they would lie with characteristic patience inside the bar, until a favourable wind should enable them to make the ascent to Marudi, sixty miles up stream; during the North-east monsoon they were about the only source of supplies to the up-country tribes. To-day steam power has improved matters; but it is still John

Chinaman, business-like, precise, and well-groomed, the "Hidalgo of the Eastern seas," as Stevenson calls him, who does the best trade on the river, carrying up supplies of food, and returning, after patient and honest barter, with rattans, gutta-percha, and other forest products.

Our journey up-stream is (or rather, was) made in the Government paddle-steamer "Adeh," a boat not without a history of her own. She was built in 1884, a vessel of 400 tons, and figured honourably on the Baram and between Baram mouth and the capital, Kuching, until 1928, when she was broken up. For many years her captain was a fine old Malay ex-pirate, named Leman, who had been captured on an expedition against the notorious Illanun pirates and was given a job of work on one of the Rajah's vessels. He was a fine navigator and crossed the Baram bar in all weathers; his merit won for him the highest post in the Rajah's fleet. He was, unhappily, drowned by reason of the capsizing of the boat in which he was taking the depth of the water on the bar, and the Sarawak Government lost a valuable and faithful old servant of many years' standing. He was succeeded by his son, Bandong, who retired from the service some ten years ago.

The engineer of this comfortable and efficient craft was the late John Mathie, of Saltcoats, who became Chief Engineer to the Sarawak Government and in the course of his official duties brought out from England a number of vessels of various tonnage for the use of the Rajah's Government.

During the first part of the voyage the country is, for Borneo, tame. The chief feature in these lower reaches is the insect-life, some interesting to the point of fascination; there are also others. The stranger gains a peculiar pleasure from the presence of a novel kind of humble-bee, not black and yellow like our English variety, but with light-blue stripes on their dark bodies. There is also another kind of bee, almost too genial to be pleasant, who, when one lies at anchor at meal-times, comes into your boat and brings his friends with him. They wish you no harm and will do you none, unless you disturb them; if left alone, they wander harmlessly over your face, licking up the perspiration. Much good may it do them!

Another curious bee is a tiny member of the

humble-bee family, about an eighth of an inch in size and known as *Kalulut*. This bee makes its home in the inside of a hollow tree, but keeps in touch with the outer world by means of a trumpet-shaped tunnel about the size of a man's finger. The ordinany passer-by is not likely to notice it, but others, who are better acquainted with the jungle, look out for it, and "when found, make a note of." The bear is fond of it, for it contains a certain amount of honey; and the Iban uses the wax of which it is made for various household purposes, and, among other things, for blocking the air-holes of the *Keluri,* a native wood-wind instrument. However, the bees do not seem to trouble much, for as soon as one is taken away they build up another.

Less pleasant creatures appear at eventide, large gadflies, rather like the pestilential horse-flies of the Alps, which turn up from nowhere in particular, and at once set to work on any unprotected part of the human body. The only interesting thing about these tormentors I was ever able to find, was that under each of their first pair of wings (as one might say, their arm-pits) they have a peculiar "balancer," like a rudimentary eye, which they use to preserve their equilibrium when in flight. When this nerve is removed, the fly loses its orientation, and is rendered more or less incapable of effective movement. This knowledge may perhaps prove of some comfort to the tortured traveller; but practically I fear it may be of no more value than the laborious method of destroying black beetles invented by a German scientist, who, when asked " Would not treading on them do as well? " replied, " That *also* is a very good method! "

Mosquitoes are as great a nuisance here as elsewhere, swarming among the Nipah palms, and, unless one has mosquito-curtains, they, together with sand-flies, painted flies, and other insects, make night intolerable. For the rest, there is but little life to attract the traveller; there is a sameness about the long reaches, except for an occasional pigeon flying overhead or a crocodile lying on the bank more or less like a log. Occasionally we meet parties coming down-stream or others making their way up, and receive and impart the latest news. When we want to cook a meal we have to find a spot on the bank where the scrub has been cleared—probably for a purpose similar to our own. To get there in safety is, however, not easy for a European who is hampered with shoes; and he is only too glad of a helping hand from his surer-footed Eastern brother.

On either bank we notice small plantations and fruit gardens, generally belonging to Malays or Ibans, and very often derelict, the owners having moved up-stream when the land got worked out. Here, especially during the fruit season, one may come across a herd of wild pigs crossing the river in search of food, and (contrary to general belief) swimming as well as any other animal. The common wild-pig (*Sus barbatus*) is found all over the island, especially near rivers, and gives good sport to those of the native peoples who are not debarred by their religion from eating pork. He is hunted with spears and packs of dogs, and on account of his protective colouring, which is of a neutral tint and blends with the colours of the underwoods, he is not " anybody's money." He has a better claim than any other beast to the title " King of the Jungle," for he fears neither beast nor man.

Two other species of Wild Pig are found in Sarawak, *Sus vittatus,* the collared or Branded Pig, so called from the white streaks running from the sides of the face to the neck; and *Sus verrucosus,* the Warty Pig, which has two pairs of large hairy warts on the sides of its face. Dr. Beccari points out that similarity of these protuberances to the expansions on the face of the *Maias,* or Orang-Utan, and the hump on Indian cattle. They probably serve as a kind of armour, in case of fights between rivals. In all the Bornean species the coloration of the young differs from that of the mature animal; the youngsters are striped with rusty red, but the stripes disappear as soon as they attain maturity. A curious fact noted by Mr Shelford is that in the young pig the upper lip is deeply indented so as to accommodate the tushes, which have not yet appeared. " We must suppose," he writes, " that the notching of the lip was originally brought about by the hypertrophy of the tusks, whereas the notches now antedate their original cause by several months."

In Celebes, and on the island of Bouru, but not in Borneo, is found that grotesque pig, the Babirusa, the Malay word meaning " Pig deer," although the animal is no relation whatever to the deer. It is a slim creature, standing about two feet at the shoulder, with a dark almost naked slate-grey skin and light-coloured eyes, and blessed with two sets of protruding tusks, one pair like those of an ordinary Wild Boar, and the other curving back so as almost to meet the forehead. Babi-rusa resemble other wild pig in their habits, going about in herds and frequenting the dense forest and riverine tracts, for they are powerful swimmers.

About sixty miles from the river mouth one comes to Baram Fort, built in 1889, in an angle between two long reaches of the river, at the village of Marudi, now known as Claudetown, after Mr. Claude Champion de Crespigny, the first Resident of the Division. The Fort is built on a low flat hill, and contains a Court House, Police Station, the barracks of the Sarawak Rangers, and comfortable officers' quarters with the luxury of a billiard table. The Rangers and Police lines give the place a military appearance, which is enhanced by the presence of a couple of elderly cannon, seconded for civil duty.

The Residency is a comfortable, well-built house, with a really beautiful garden of flowers and flowering trees, both indigenous and imported. There are a number of pergolas covered with all kinds of lovely creepers. Bougainvillea in two shades, pink and magenta; the bell-flowered Ipomoea, or Morning Glory, a deep bright blue; Plumbago running wild; two kinds of Bauhinnea, with prolific blossoms of gold and crimson. Elsewhere there are moonflowers, tuberose, caladium with their blotched foliage, allamanda, oleanders and the waxen-flowered ixora, both the white and the red. Ferns and orchids (especially the Vandas) are everywhere; while one of the most striking of the local plants is the Anæctochylus, which is found in the thick moss growing on mountain tops, its velvety leaves veined with gold or silver. Among the imported flowers which thrive, are the Stephanotis from Hawaii, and the beautiful Spathodea, which was introduced here many years ago from Uganda. This tree, which grows to about the size of an oak, bursts into blossom with large clusters of bright red and orange flowers, which stand out boldly against the dark forest background. For the local children it provides a fascinating toy, for the flower-pods, before blossoming, accumulate small quantities of water and provide them with a cheap water-pistol or squirt. Effective use can be made of these, as I can personally testify.

In the bungalow itself, Man is by no means the only inhabitant, a number of harmless, if sometimes vexatious, creatures cohabiting with him. Perhaps the most pleasant are the lizards, of which there are many, large and small; the commonest is the *chic-chac*—well known all over tropical Asia, and so called from his chirruping note. During the day he stores himself away in some cool dark retreat, especially at the back of pictures, but at night, when the lights attract various insects, he shows himself on the walls, displaying such enterprise and agility in his *shikar* that his colour gradually changes from a transparent buff to a rich brown by reason of the insects he has swallowed but not assimilated. A larger lizard, perhaps eight inches in length, is the *chocwah,* who occupies his business in rafters and beams, emitting occasional and startling noises, and perhaps flopping down unexpectedly on to your table.

If one lives *en garçon* in the East, one becomes perhaps careless, at least as regards domestic tidiness, but when the mistress of the house arrives, she clears up a number of things which she (very properly) considers eyesores, but may have endeared themselves to you by their very familiarity. Among these are the works of the Mason-wasp. This very delightful and really harmless insect seems to have a special affection for human habitations; it has a habit of plastering your furniture with comical blobs of mud, from half an inch to a couple of inches across; these are its nests, in which it lays its eggs and brings up its family. The cells are like the craters of miniature volcanoes, and are run up with extraordinary rapidity, the builder bringing in a fair-sized piece of mud every two or three minutes. If allowed to complete her work, the mother wasp, like Browning's " Small Man," " goes on adding one to one," until she has fifteen or twenty of them. Very often she adds a sort of external decoration, consisting of streaks of mud of a different colour. Her object seems to be to disguise her nursery by making it look like a piece of tree-bark.

This wasp is continually pestered by parasites, which, cuckoo-like, prevent the development of the larva by laying their eggs in the cells and so giving their offspring the benefit of the spiders and other creatures which the wasp has carefully placed there for her own children.

Another hard-working and persevering insect is the (so-called) Carpenter Bee, a large and beautiful insect which bores holes about the size of a man's little finger in any soft wood—the roof-beams of your bungalow for choice—scattering chips and dust all over the floor. If you try to get rid of it, by sending someone up a ladder to plug the holes, the bee simply bores through the beam, singing and humming as it works, and, having made its way out, starts the same business somewhere else. This bee has a sort of country-cousin, which lives in the forest and is especially fond of hollow (or so-called female) bamboos; they are both diabolically persistent, and if once your house has become their home, you will have no peace until you have removed all the soft wood and replaced it with a kind of hardwood which they like less.

Perhaps, however, the worst are the ants, which swarm everywhere and find an unholy joy in devouring one's pet specimens. It is even necessary, at times, to keep one's sideboard or food-safe standing in little cups of paraffin in order to keep off these pests; but even so I have known them to bridge the interval with pieces of grass, straw or small twigs, in order to get to the meat in the safe. Cockroaches at times are a great nuisance, for, being able to fly, they manage to get at one's food in quite unheard of ways.

When darkness comes, the lights of the house, especially during the rains, attract all sorts of flying creatures, including the harmless mole-crickets, which are more or less blind, but come in with the rest, striking one in the face or settling on any dark cloth. It is a very curious fact, noted by Dr. Russel Wallace and confirmed by others, that moths and other insects are not attracted by light unless the light has been shown for many nights.

There is one of these night insects, however, which it is a delight to watch; this is the well-known fire-fly, whose light has an almost supernatural beauty. No one has yet explained the function of this light; but Mr. H. N. Ridley suggests that it serves as a defence or warning, having seen a small-sized lizard frightened away by the sudden flashing of it. On the other hand, it is quite likely that it may be used for purposes of recognition, since the females flash at a different rate from the males. The late Dr. Willy Kükenthal, the Jena naturalist, who stayed with me at Claudetown, perhaps thought of fire-flies with mixed feelings, for being much troubled one night by mosquitoes (who had turned his body " into one great wound ") he cursed them for having called in the assistance of the fire-flies to light them on their way. As a matter of fact, mosquitoes prefer the dark, as one knows when one has one's feet under the dinner-table. In Sarawak a sheet of notepaper under the socks is found a useful protection; but I have heard of houses in India where pillow-cases were regularly provided as a protection at dinner.

CHAPTER IV

THE UPPER BARAM AND RIVER-FOLK

The Real "Dayak"—The Tinjar River—Native Methods of Fishing—The Magician Fish—Crocodiles and Crocodile-fishing—A Battle Royal—Turtles—A Tortoise Story—Herons—Wild Cattle—Two Wonderful Butterflies—A Deer-hunt—Native Compassion—The Water of Life—The Mystery of the Forest—Flying and Gliding Animals—The "Gin and Bitters" Insect—Flying Foxes—Phosphorescence—The Monkey and the Toad.

It is above Baram Fort that the real jungle begins, for here the primeval jungle takes the place of marshland and of secondary forest, *i.e.* forest in which partial clearing has taken place or trees have been felled; for where there has been clearing or cultivation, one finds, not so much the representative flora of the island, but species which may have a wide geographical distribution; moreover, the type that we now meet in the river-side villages has changed. The Chinaman and the Malay of the coasts have disappeared, and with him the trim little gardens with their strange native vegetables, ladies' fingers, Aubergines, and the Lord knows what else. We are in the land of the real "Dayak," the up-country man, the Kenyah, and the Kayan, the last of whom, among the other pagan tribes, is almost considered a foreigner, but exercises a paramount influence. Here we find the typical Long House, perhaps three hundred or more yards in length and containing eight hundred to a thousand souls. These are commonly built above flood-level and especially at the junction of two rivers, the spot being chosen for strategic purposes; although it is possible that the Bornean aboriginals have, like the natives of India, a religious sentiment or superstition about such a position.

After a time we are joined by the Tinjar, the chief tributary of the Baram, which is of sufficient length and volume to be considered a sister-stream, and, after another twenty miles, by the Tutau, which enters the main stream on the other side; about sixty more miles, however, still remain before we abandon the launch and take to the local canoe, manned by Kenyahs, a great river-people. Our new vessel is propelled by the paddles of some stout

A TYPICAL SARAWAK JUNGLE SCENE

fellows, who, when we come to rapids, will either pole us up stream or else, leaving the ship, help us along by hand. For the river has now narrowed considerably and in the dry season presents long stretches of shingly beach, or else, where there is a slope, has broken into a set of narrow currents rushing with great force between rocks.

Between such reaches, however, we come upon spots where, by a sort of natural dam, broad lakes have accumulated. In the cool shaded waters among the tall rushes and other water-plants congregate quite a number of fish, which find here a pleasant shelter from the sun's rays. Some of these are peculiar to Borneo, among them a curious kind of Toad-fish known locally as *Kittabu* (*Batrachoides*). In the lakes at the side of the river, but, curiously, not in the river itself, may be found a most uncommon fish, known locally at *Silu* (*Scleropages*).

The native method of fishing these waters has the merit of simplicity, and has apparently been in vogue for centuries. The chief part in the business is taken by the root of a plant called Tuba (*Derris elliptica*) which stupefies the fish. Large masses of this sort are collected at a likely spot—such as a riverine lake, or a spot where the water has run low—and distributed among the boats of the prospective fishermen. The boats are now partly filled with water, in which the root is pounded and rinsed until all the juice has been extracted, and the water turns milky. Then at a given signal all the boats are upset into the enclosed space of water; the fish quickly become aware of the presence of the poison and rush wildly to and fro, only to find their escape barred by extemporised platforms like wharves. At this point the excitement begins, for the fish leap in hundreds on to the barrier and are either speared or clubbed by the sportsmen, who often have to be very nippy in order to avoid the spikes and thorns with which many of the fish are armed.

As a sport, the fishing is not much better than making a slam at Bridge when you have all the cards; but there are certain superstitions and folk-stories connected with it which are far from being banal. Natives, for instance, when making preparations for this form of sport will never mention the words for fish or the root, for they believe that if they did so, birds and other flying creatures would warn the fish of the plot, or would give information to a certain Magician-Fish, called by Kenyahs Belira, which would be able to call down rain from the sky and so, by swelling the volume of water, prevent the successful polluting of the stream. This Belira is a large flattish fish containing a stupendous number of bones, the presence of which is accounted for by the following story.

In view of the persistent contamination of their waters by Tuba-root, the fish of Borneo in Council assembled:

> *Resolved* That a Dayong or Magician be forthwith appointed, with power to call down rain from the sky and prevent said contamination. Applications for the post to be submitted to the Chairman.

The applicants were not numerous, the duties being arduous and the remuneration problematical; but at last the Belira fish offered his services, conditionally on the payment of one bone from each fish. The tender was accepted, and the payment duly made; but the Belira seems hardly to have acted up to his promise.

Another story recounting how the Animals of the Jungle, observing Man's success, resolved upon a tuba-fishing of their own, the trituration of the root to be performed by ordinary chewing. Unhappily the only four-footed creatures who could deal with *Derris elliptica* without harm to the system, were the Porcupine and the Rhinoceros, who thereupon set to work. The Porcupine did his bit, and the Rhinoceros (whom legend credits with a somewhat pompous humour) chewed Tuba-root for twenty-four hours at a stretch, finding perhaps that it promoted a healthy action of the skin. The two then proceeded to the river-bank and discharged the juice, much to the consternation of the fish but to the satisfaction of the other members of the Animal Kingdom, which was thus able to prove its superiority.

Pollution by the Derris root is, however, not the only terror of these placid waters; for near the water lurks that noisome pest, the man-eating crocodile (*Crocodilus porosus*) waiting for animals who come down to drink. A very short shrift is the victim's; I once saw a crocodile seize a good-sized pig and disappear with it under water. Perhaps two minutes later, half the pig floated to the surface. In these circumstances it is hardly to be wondered at that swimming in the open river is not popular among Europeans, even in the heat of the day. The best that one can do is to confine one's self to the shallow water near the edge, with a boat moored a short distance out, in which natives sit, banging with their paddles on the side to keep the brutes away.

This crocodile is not only very fierce, but very large, often measuring twenty-three feet or so in length. Its skin in clear water is of a greenish-black and yellow, but when it reaches the swamps and muddy banks lower down the river it accumulates on its skin such an amount of mud and slime that it looks very much like a log and often is mistaken for one by its innocent prey. Its eggs are prolate, *i.e.* longer than they are broad, and measure about three and a half inches in length; the shell is pure white but semi-transparent like delicate porcelain. The female lays thirty or forty at a time in a depression scooped out of the mud, where grass and herbage can be found.

A rarer species occasionally met with in Borneo

is the Gavial (*Tomistoma schlegelii*), which has a long snout with a big knob on it; as far as I know, it does not eat human being, but restricts its activities to fish and the soft-shelled turtle.

When fishing for crocodiles the natives use a line about sixty feet long made of rattan, with a swivel bar of wood two or three feet long and pointed at each end, fastened at the end of it, like a toggle. The bar moves easily in any direction and is intended to catch the inside cheeks of the crocodile, for whose further discomfiture the last few feet of the rattan rope are frayed, so that they may entangle his teeth without being completely severed. The bait consists of the body of a chicken or of a small monkey and floats on the surface of the water, being fastened to the centre of the bar which is the business end of this ingenious contrivance, the other end being fastened to a floating log; for a crocodile will not take a fixed bait. The anglers watch from a boat at a convenient distance.

I once assisted at a catch and found it sufficiently exciting. Watchers had told us that a crocodile was " on ", and four Malays and myself pushed off in a dug-out canoe, taking on board the rattan rope and also carrying with us a coil of ordinary rope and a sort of harpoon. As we shortened our line, the crocodile felt us and the rattan moved violently. We at once gave it a good jerk so as to catch the creature with the points of the toggle. For a second or so there was a lull, for evidently the crocodile was " on " and was temporarily bewildered; and we began to pull in the line. Suddenly, however, he dived, realising what had happened and, trying to break us, started to rush madly about, towing us after him. We at once started shortening our line, expecting every second to be capsized, and after a confused scramble came up to him. We continued to haul in the line until the crocodile's head came out of the water, when I stabbed with the harpoon at its throat. The spear entered up to the socket, but the great beast clutched at our gunwale with its taloned claws and see-sawed us to and fro like the escapement of a watch, we meanwhile letting him have more rope. As we did so, the crocodile got entangled in the two lines, and, pulling in, we soon had him again within a few feet of us. Acting on the instructions of our leader, I held tight with the harpoon line and hauled in on the other, while my companions, as the great open jaws appeared above the surface, deftly slipped a noose over them and closed his mouth; for it is known that if a crocodile cannot get his mouth open he loses heart. All went now according to plan and we were able to bind more cords round his jaws, and then to make

SOFT-SHELLED RIVER TURTLE

fast his forelegs over his back; in which undignified position he was towed to shore and there decapitated.

Around the shores of these lakes, in crevices between the roots of bushes and trees, one may see various fresh-water turtles and tortoises, especially the Soft-shelled Turtle, known as *Labi-labi*

(*Trionyx subplanus*). This creature has a quaint, flexible snout, rather like a miniature trunk; and, curiously, though soft-shelled itself, lays a hard-shelled egg. Here also, basking in the sun and almost indistinguishable from the mud, may be found that interesting tortoise, the *Tekora* or *Kelap* (*Cyclemys spinosa*), whose carapace is blackish-red and ornamented with curious spines growing out from the scales and along the rim, while the plastron, or under surface, is flat and mottled with pretty yellow and black markings.

This tortoise is the subject of many popular stories, like those of Brer Terrapin in "Uncle Remus." According to one of them, he and the Plandok, or Mouse-deer (*Tragulus*), once went out in search of fruit. Finding a quantity in a garden near a village they helped themselves, the Tortoise doing all the hard work and climbing the trees; and were dividing the spoils when the villagers discovered them, being roused by the noise made by the Plandok, who kept calling out in a loud voice, "This is for you; that is for me," as he dealt out the fruit. The Plandok, taking as much as he could carry, escaped, as he always does in these fables;

the Tortoise hid for a time under some large leaves, but was at last found.

"This is the thief," cried the villagers. "Come, let us kill him!"

"Put him in the fire!" suggested one.

"By all means!" said the Tortoise, "The last time I was not baked properly," and he showed his light-coloured plastron.

"That won't do, then," said the villagers. "We will put him in the sugar-press."

"I wish you would," replied the Tortoise. "Only, do the job thoroughly. Last time you left it half-done," and he indicated his domed carapace.

"Likes it, does he?" said the villagers. "Well, then, throw him in the river! See how he likes that!"

"No, no!" cried the Tortoise in tears. "Water is very bad for my health."

"That's the stuff to give him!" said the natives, and they threw him into the river; whereupon the Tortoise, when he was at a safe distance, called out: "Many thanks! This is where I live! Good-byee!" and, waving his flipper, he disappeared.

The fish which abound in these riverine lakes

attract a number of interesting and beautiful birds, such as the well-known Adjutant Bird, and the Indian Darter, that curious serpent-like, kink-necked bird which Mr. Shelford has described so well; also various beautiful species of herons. The large blue heron may be occasionally met with here, but as a rule he frequents the seashore and river mouths, where pairs are often found fishing in the shallow coastal waters, for which they have temporarily deserted their nests high up in the branches of the tallest trees. Two other species are, however, fairly common: a brownish-grey Heron known locally as *Kuju,* which has habits similar to those of the Blue Heron, and lays eggs of a dull blue colour; and the smaller Night Heron, distinguished by two long slender feathers of a very light grey springing from the crown of its head. These birds build in colonies, like our rooks at home, in the tops of trees, and make their nests of twigs; the eggs are bluish.

A peculiar bird-note that one may hear in these localities is the whistling made in flight by the wings of a bird of the Ibis family, known as *Burong Undan.* This is a large black bird, whose feathers are shot with a metallic purple; as it usually flies up above the forest trees, it is more often heard than seen, a fact which perhaps gives rise to many native legends about it, for it is roundly asserted that the bird is a phantom, even more than Wordsworth's cuckoo, "an invisible thing, a voice, a mystery."

Many striking water-plants grow in the mud of these lakes, among them a variety of the family series of lilies known as Crinum (*Amaryllidaceae*). These lilies have long broad leaves, some as much as eleven feet in length; they grow close together in the water and often cover as much as an acre. Growing on the stems of these plants you will find the beautiful *Vanda* orchid named after Sir Joseph Hooker, a lovely flower of pink and white.

As we ascend, leaving the lake and the river, the forest growth becomes denser and approaches closer to the banks, except for small clearings, which, once the home of man and the site of extinct villages, are now inhabited by the wild cattle known to the Malays as *Banteng* (*Bos sondaicus*). This, the largest animal in Borneo after the Rhinoceros, is a handsome creature but very shy. One does not often

FROG-MOUTH FROM THE FOOT-HILLS
(Batrachostomus harterti)

HOSE'S BLACK ORIOLE FROM MOSS-CLAD HILL-TOPS (oriolus hosei)

come across it in the flesh, but sees its traces, especially where it has rubbed off the mud on its back against the trunks of trees, about four feet from the ground. It prefers the secondary to the primeval jungle, one of its favourite articles of diet being young bamboos.

The river now becomes shallower but wider, and a gravel bottom necessitates constant poling; the river-bed is everywhere studded with small boulders, over which the river ripples musically, an attraction to innumerable dragon-flies. These are beautiful iridescent creatures, slightly smaller than those that we see at home, and they flit about over the surface of the main stream or settle in the pools and half-submerged pebbles in the shade of the bank, in their constant search for smaller insects. In these shady spots one may, if one is lucky, catch a glimpse of that beautiful butterfly, *Ornithoptera brookeana,* named by Dr. Russel Wallace after the first Rajah of Sarawak.

This is a large insect of velvet-like green and black on the upper wings, and with two bright red spots on the head. I once shot one of these butter-flies with a rifle—a feat which sounds Transatlantic until explained. I was doing some target practice with a small-bore rifle; when we went up to the target to examine our hits, we found that one of my shots, in transit, had struck an Ornithoptera, which was lying on the path with wings flapping. All the same, the butterfly looks big enough to shoot, for he is larger than some of the Bornean birds.

Another beautiful butterfly is the one called *Hestia idea,* which is about five inches across the wings, and whose large grey and white wings make it look like a miniature sea-bird. It gives the impression of always being about to alight, but never doing so; as it sways up and down in a fascinating indecision, its sinuous sweeps show to the best advantage the almost transparent gauze of its wings. One almost feels that the creature is aware of its own beauty, and that it times its flight so as to provide the spectator with a natural slow motion picture.

It is a curious fact that there has been found in Java and in British North Borneo, a very rare butterfly (a new sub-species of *Elymnias künstleri borneensis*) which is said to mimic this butterfly. I am hardly inclined to believe that such can be the case, for although the new sub-species may be a mimic as regards colour, shape, and size, I cannot imagine that it can imitate the flight of *Hestia*. If it does so, it is one of the most remarkable cases of mimicry in Nature.

Elymnias künstleri borneensis,
the butterfly which mimics *Hestia ideæ*.
(By the courtesy of Capt. Riley of the British Museum)

On one such journey as this, I remember, as we turned a bend of the stream, I noticed some empty boats tied up to the farther bank, and heard the barking of dogs and shouts of men; suddenly there darted out from the forest a fine old stag, closely pursued by a pack of the nondescript dogs that Borneans keep for hunting purposes. Leaping into the stream he swam powerfully in our direction and soon was well away from the dogs; but now four men arrived in pursuit, two making down-stream to where their boats were tied, while the others ran up the stream towards a spot where a number of small rocks and ridges of pebbles made crossing easy. Evidently their aim was to cut off the stag in mid-stream before it could reach the bank; and the men in my boat wished to take part in the sport, for we were lying close in under the bank for which the animal was making. With some difficulty I restrained them, for interference seemed to me unfair to the stag; but we enjoyed an exciting spectacle. The stag made good headway against the rapid current, but the two men with the boat had put off with extraordinary rapidity, and were now only a yard or so behind him, while the two upstream, taking flying leaps over the gravel and rocks, had almost reached the bank, and seemed certain to cut off his escape. For a moment or so it was anyone's game. The stag dashed from the water on to the stony beach, and almost in the same second the boat grounded, upsetting the balance of the man in the bows. With one leap the stag reached the top of the bank, the foremost man making a slash at him with his *parang,* but the blow fell short. At the same instant one of the men up-stream shouted and threw a spear, which missed. The stag had won, and bounding off into the dense jungle was lost, much to my pleasure, but to the annoyance of the sportsmen. Cursing their luck, and blaming each other's incompetence, they threw themselves down on the bank and consoled themselves with local-made cigarettes. They could not understand why we should have held back, when it would have been so easy to capture the animal for them while it was in the water. They shouted most vociferously and gave a general display of temper, which I fear I did not improve by remarking that there were plenty of other stags in the forest.

I am a Norfolk man and remember how George Borrow's friend, Jasper Petulengro, refused to give the Romany Chal an unfair advantage over the Plastramengro. For the same reason I was sorry when I read the other day, how in Somersetshire, of all places in the world, where men are supposed to be sportsmen, servants had been sent out in a boat to cut the throat of a stag which had taken refuge in the sea.

It must not be thought, however, that Borneans are insensible to animal suffering. Many are quite the reverse, as I learnt in a Kayan village in this very part of the country. As we paddled slowly up the stream, we came across a man accompanied by some dogs; he had been hunting and was now returning home. Recognising me, he asked if I would like to have a little mouse-deer which his dogs had chased into the river and he had captured unhurt.

"Yes," I replied, and gave him a pound of native tobacco for it. As he lifted it into my boat I noticed that she was heavy in young; her feet were tied fast with bark-string, and she seemed at the point of death. Soon, coming to a spot where the bank was less steep, I told my crew to stop, and with the assistance of one of them untied the lashings round her feet.

The poor little thing must have thought that her last moment had come and cried and moaned piteously, stiffening herself in such a manner that for a moment I really thought she had died in my arms from sheer fright. I carried her ashore and gently laid her down near the top of the bank. She got up, ran a few feet towards the forest and then stopped, deliberately turned round, and stared hard at me with her large glistening brown eyes.

The boatman who came on shore with me remarked: "You have let it go; perhaps it is the Tuan's spirit helper."

The frightened little animal quite unexpectedly stopped to look at me for a moment instead of making for safety as quickly as possible directly she had a chance to escape.

A few minutes later we reached the Kayan village and the men commenced carrying up my belongings to the house. The story had, however, preceded me, for after sitting chatting for a while with the chief on the reception dais, he invited me to his room, where I found his wife and a party of her friends ready to welcome me; and we sat down for a smoke and a talk.

Suddenly the Chief's sister, an exceptionally nice woman whom I had known for several years, apparently unable to contain herself any longer, asked me to tell them the story of the mother mouse-deer which I had let go free.

I replied that I had acted out of pity for the poor creature and that I hoped it would escape safely. A slight murmur of applause went round the circle, and Oyong Obong remarked that Kayan women had the same feeling of pity (*Masi*) for all creatures heavy with young. "You will be rewarded," said she, "for your kindness."

The Mouse-deer, the Muntjac or Barking Deer, and the Rusa are three of the commonest deer in Borneo, although they are not found exclusively in the island. The Rusa is a variety of the Indian Sambhur, only with smaller horns—six points usually. It measures about fifty-two inches at the shoulder and is of an almost uniform dark brown colour throughout. It is a timid animal and often a solitary one; its home is the jungle, but it comes out on to grassy slopes and clearings to graze, usually at night. Very often they take to the hills and may be found in the entrances to the limestone caves, seeking shelter and coolness from the mid-day sun. The flesh, which is eaten by some natives, is always cooked in the open for it is thought that if it were brought indoors its timid spirit might be infused into the inhabitants.

It was on a journey once in this part of the river that my Kayan boatmen suddenly asked to be allowed to halt, as they wished to bathe in a small fall of water which oozed from a black rock at the side of the river. I gave them leave, and they stood under the wall, letting the water fall all over them and splashing themselves with it, asking me to do the like.

"Why?" asked I; and they replied that this was *Telang Ading,* the Water of Life, and told the following story.

"Many years ago, so our fathers say, there lived near here a huge boar with long tusks and with whiskers on his cheeks. Our fathers pursued him for years, but he could never be caught, for he ran like a deer. At last, one day, they found him in the stream. He fought like a demon, killing dogs and biting iron spears in half; but at last he fell and they left him for dead, and went off after another pig. When they came back to fetch the body, it was gone; and all they found was the traces of his feet. But many months after, he was seen again higher up the river, as strong as ever. How could this be? Then they remembered that he had been left near this waterfall; and there is virtue in the water, for it can bring the dead back to life. Therefore when we pass we never fail to stop, even if we are hurrying; and all Kayans and Kenyahs do the like, for the water is the water of life."

As evening approaches, one feels more and more the superhuman majesty of the forest. The sea is not more mysterious, more manifestly infinite; but the forest has, beside this, an overmastering personality, so that one feels that one is being constantly watched by some resistless monster, the eternal silence broken only by sounds which seem to intensify it. It is at twilight that one feels the immanence most strongly; and one hardly wonders that the pagan inhabitants should think of the jungle as the home of unknown powers. Before the day dies, however, certain twilight animals come out, among them the graceful flying squirrel (*Pteromys nitidus*), setting out from its hole in a hollow tree in search of food. In colour it is a light chestnut, inclining to black, with a long black tip at the end of its bushy tail; the under part of the body is a light yellowish red. It is about eighteen inches long, and has a tail of about the same length; but its chief feature is a flap of skin like a parachute which extends from the outside of the fore limbs along the sides of the body to the hind legs, and is stiffened by a spur-like cartilage springing from the wrist. This spur, when the animal is at rest, lies almost flat along the bone of the fore-arm; but it can be spread out at will, and then projects at almost a right angle to the wrist, drawing the membrane taut and giving it considerable outward extension and support. This parachute enables the squirrel to glide for a distance of eighty yards or more without touching the ground; but it always does so from a higher to a lower level.

Another gliding animal to be found in Sarawak is the so-called Flying "Lemur," or *Cobego* (*Galeopithecus volans*), whose habits are similar to those of the Flying Squirrel. This is a small animal about the size of the domestic cat, greenish-grey in colour, and provided by nature with a parachute-like membranes somewhat similar to that of the Flying Squirrel, but less effective; by means of it the "lemur" is able to make flying leaps from tree to tree—a most useful talent, for on the ground it is more or less helpless, supporting itself with difficulty on its fore-arms and shins. It only comes out into the open at night time; during the day it hides among branches or on the bare grey stems of dead trees. Its protective colouring of greenish grey and brown mottled fur renders it easily mistaken for a knot in the trunk or the broken joint of a lost branch. Here it passes its days clinging by all four feet, like a sloth; and you may perhaps find a mother in this position, with a baby clinging to her breast, and between her and the branch.

Borneo is fortunate in the possession of two other flying or gliding creatures, a lizard and a frog. The lizard, which is a kind of *Gecko* (*Ptychozoon*) is an ugly but striking-looking creature with two extensions, one on each side of the body, like the brim of a Trilby hat. It has a curious fringed tail which ends in a flat spatula, like the paddle of a canoe. The frog is rather larger than a full-grown English frog. The whole of the space between the long toes is occupied by membranes, the area of which, when the toes are extended, is larger than the whole body of the creature. When it glides, the limbs are held close to the body so as to form one continuous surface; the direction of the glide can be altered by

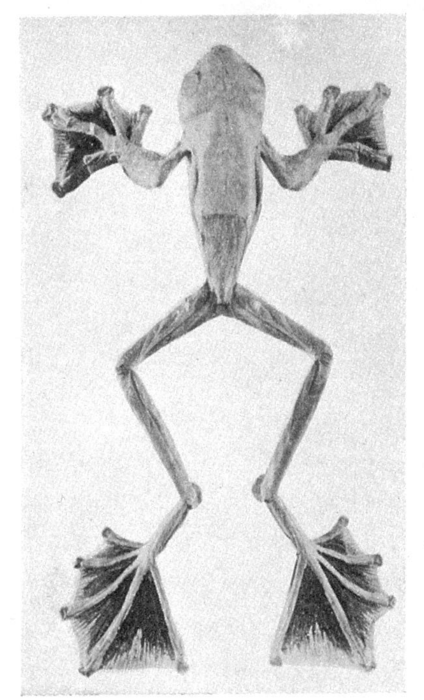

(Upper) GAVIAL (Tomistoma schlegelii)
(Lower) FLYING FROG (Rhacopterus nigropalmatus)

powerful strokes of the hind legs. This frog cannot do long flights, but he can glide very much farther than an ordinary frog can jump, and his landing is much smoother. There are also one or two varieties of snake which are able to glide from a higher to a lower level, and Captain Flower, in "The Procedings of the Zoological Society, 1899," reports that he saw a small specimen take a flying leap downwards and outwards from an upstairs window on to a tree, from which it then crawled away among the foliage.

Camping for the night, in a part of the country where one is known, is an easy matter; for if one does not stay in one of the local Long Houses, one's followers very quickly run up a shelter of *kajang,* and, before one knows it, one finds a fire lit and dinner being actively prepared. On this occasion in the few minutes that I had to wait, I was sitting idly smoking on a log, when, glancing down, I became aware of a bird sitting on her nest, almost under my feet. It was a fire-backed pheasant. Quite undismayed, she sat close, just like our English bird; I made no effort to disturb her, but returned to the spot next morning. She was still on her nest, but as I approached, she hopped off into the forest. In the nest were several eggs, like the brown eggs of a domestic fowl, but marked with small spots. These I left undisturbed, and I have no doubt they hatched out successfully.

Sunset and nightfall, which are almost simultaneous, is heralded with extraordinary punctuality, by an insect orchestra, the leader of whom, to judge by the noise, is a creature called Ingit, but known to Europeans as the " Gin-and-bitters Insect." Not the least of these musicians are the cicadas, creatures beloved of the Athenians in the fifth century, and here in Borneo the Far East, arrayed in gorgeous livery of electric blue, green, black, and grey. Their famous song, about which the poets have, one thinks, used a certain license, is produced by a sort of flexible notched rib under the second wing, which makes an interminable noise like a circular saw. If the circle sings " like a King," as Anacreon says, I can sympathise with Greek tyrannicide.

With the insects appear numbers of bats of all kinds, too many to mention here; among them is the " Flying Fox " (*Pteropus edulis*)—which is not a fox at all—a large fruit-eating bat, whole companies of which may be observed hanging down from the branches of trees and looking like a collection of half-open umbrellas. Then there is *Cheiromeles semi-torquatus,* which is almost hairless and has a most unpleasant odour, rather like burnt leather. It lives in holes in the great Tapang trees and has thick leathery wings, with membranes attached to the sides of the body, upper arms, and thighs, in such a way as to form large pockets. The female uses these pouches to carry its young in; but curiously, as Mr Shelford notes, the males are similarly endowed. One therefore wonders how, and why, it came about. A curious fact about both these bats is a parasite which lives either on or with them; this is a sort of wingless fly, the pupa of which is often found embedded in the wing membranes. Even more remarkable is a strange earwig (*Arixenia esau*) which lives in the brood-pouches of this Cheiromeles bat.

Other interesting bats are the Leaf-nosed Bats (one species of which is found in the British Isles), in which the muzzle is surmounted by a leaf-shaped appendage. A large blue-grey species, although apparently less strong on the wing than most bats, is yet said to prey upon smaller bats, and small birds, as well as on frogs and insects. Another species, the Long-eared Bat, puts himself to bed, as it were, by numbers, first folding up its wings and tucking them close to its sides, and then doing the same with its ears, one by one. Like other bats of the same class, it sleeps head downwards.

A feature of the Bornean forest at night is its phosphorescence. Flying insects spangle the dark background with countless fairy lights, some stationary, some moving in graceful curves and some flashing out for the space of a second, and then dying, with all the suddenness of a shooting star. Even the leaves, dead twigs, and the very earth seem to give forth a radiance; by bamboo clumps can be seen luminous fungi so bright and so numerous that they almost light the path between the trees, turning the forest into a fairyland.

Now if you gather round the fire with your native friends, you will doubtless hear folk-stories appropriate to the time and place. An especial favourite concerns Kra, the monkey, and his friend Raong, the toad. These two creatures, one particularly rainy night, resolved to get up early next day and make themselves waterproof cloaks from the bark of a certain tree. But on the morrow when the sun shone warm and bright, the monkey went off to his usual haunts among the tree tops, while the toad climbed on to a log and basked contentedly.

"What about those cloaks?" asked the monkey at last.

"To-morrow will do," replied the toad. "Why not let well alone for to-day?"

And so it has gone on; and any night when it rains you will hear the croak of the frog and the growl of the monkey, complaining of the cold and wet, and determined to go to work to-morrow.

Reflecting upon which moral story, one goes to bed oneself, doubtless with equally good resolves.

CHAPTER V

A SHORT-LEAVE HOLIDAY

Equipment and Composition of the Party—Soul-capturing by the Camera—Volunteer Helpers—" *Pina!* "—Gibbons—The Hand of the Ape—Macaques—The Sacred Monkey—Hanuman—The " Dutchman "—Bees—Red and Black Ants—The Termite—Its Value as a Political Guide—Sun-birds—A Bird that Sews—The Drongo—Various Cuckoos—Community Singing—The Crow-pheasant—Bird Stories.

IN the north of the Kayan country, and between the Apoh river (a daughter stream of the Tutau which, in its turn, is a tributary of the Baram), and the main river itself, rises a conical hill known as Batu Song, to the height of some five thousand feet. On account of its shape, which somewhat resembles an inverted mortar, it is a well-known landmark; and by its character (for it is a sandstone hill) it gave promise of interesting flora and fauna. To this mountain I determined, taking advantage of a period of administrative and magisterial quiet, to make an expedition in order to collect specimens.

A SHORT-LEAVE HOLIDAY

I should say that, in my later years in Borneo, I became more of an observer than a collector; but this particular journey was made with the definite intention of procuring new specimens from a mountain hitherto unvisited by Europeans.

My equipment was as light as possible, for temporary shelters can be put up quickly by the forest tribesmen, who are particularly expert at this sort of work; I therefore took only bare necessities such as provisions (chiefly rice), cooking utensils, and a plentiful change of clothes. The atmosphere of the old jungle is so damp that garments can only be dried over a fire, which is often inconvenient and ever unpleasant; as for one's boots, they have to take their chance, and, although putting on wet shoes in the morning is a horrible business, one soon walks it off. Such things as one requires are easily carried by a few native helpers, among whom, on this occasion, were two Punans, members of the most unsophisticated and primitive of all the Bornean peoples, whose home is the forest, and who rely for their daily food on the use of the blowpipe and the skill with which they can imitate the noises of birds, beasts, and insects. Bornean porters can cope with quite heavy loads, up to sixty or seventy pounds per man; they carry a basket slung on the back by means of light straps made of tree-bark and fastened to a board, a third strap being carried over the head in order to lighten the burden.

An important part of my outfit was my camera. It was a half-plate stand-camera, specially made for me by Messrs Reynolds & Branson of Leeds; I used an anastigmatic lens and isochromatic dry-plates. Knowing many of the natives as well as I did, I never experienced any trouble with sitters; they took an interest in my pictures, helped me to find subjects and were highly delighted with the finished products. With strangers, the case was altered; they semed to be afraid and hid their faces, alleging that the camera was an instrument used to capture their souls. " He sets up a box on a stand," they would say; " first he moves you about, then he asks you to keep still; and before you know it, he has caught your soul and put it in the box, and you never see it any more." Another account said " The box sucks in your soul, and then spits it out onto a piece of glass." But with friends there was no lack of confidence. When they understood, they were only too ready to help; and my Milano servant, after a time, got so far as to be able to develop the plates and make prints. And very well he did it.

The expedition proper started from the junction of the Apoh river with the Tutau, where there stood a Kayan village belonging to Tama Usong, an important chief and an old friend of mine, who had promised me his assistance. Here I collected my stores and my forces. I had originally intended to employ three trained Iban hunters, on a fixed salary; but on leaving Marudi, where I engaged them, I found that my party had increased by one, another Iban having, Bornean fashion, come along for the fun of the thing and as a volunteer. I welcomed him, for an amateur of this kind is likely to prove helpful; but what was even more fortunate, was the arrival, at the Apoh, of two Punans, quite invaluable people, who not only knew the forest in general like the palms of their hands, but had actually themselves been part of the way up the mountain.

The question of suitable rewards for my three volunteers was an awkward one. To offer them cash was out of the question, for they would not value it at all, and would think that it was given merely to be returned to the Government in the form of income (or door tax). At the start of the expedition, therefore, I gave them some tobacco and some turkey-red cloth, as an earnest of good things to come. They seemed to be pleased, but rather taken aback, and proposed giving me in exchange some wild rubber; I explained that I did not wish to trade with them, but wished them to accompany me to Batu Song; which it now appeared they were only too willing to do.

Later on when, at the end of our six weeks' expedition, we returned to Tama Usong's village, I gave them a *parang,* some iron rods to bore blowpipes with, and two skeins of beads each. The latter they very much valued, for they use them both as ornaments and as currency.

" Pina! " said they (" What a lot! "); and they were much exercised as to how they should repay me. I wanted nothing from them, of course, but in the end accepted some daintily carved wooden spoons and plates.

While this was going on, my friend the volunteer Iban chimed in. " What about me? " asked he.

"I have something for you," I replied, and showed him an old-fashioned but finely chased muzzle-loader which I had with me. (For my own use, I carried a twelve-bore and a twenty-bore collector's gun with dust-shot.)

This fine old piece I promised to give him on our return to Marudi; as a matter of fact, I gave it to him shortly afterwards, when his pleasure knew no bounds.

With these six shikaris, a Chinese cook, and my own Malay body-servant, we started up a tributary of the Apoh, passing in the first day a number of Kayan villages, and very often having to drag our boats because of numerous great logs felled right across the stream. By the second day we were past such obstacles and the river had become a small stream with a pebbly bottom. On the third day we found that we could go no farther by boat, and encamped for the night, putting up a temporary shelter. Next day we left the river and our Kayan friends, choosing the early morning for "Zero hour," partly for its freshness, but partly also because then, even more than at sunset, bird and animal life is always observable, and most attractive.

Especially one notes the little Gibbons (*Hylobates mülleri*), whose "two-fold shout" sounds almost like that of a bird, as they swing from tree to tree, gathering the dew from the leaves, and seeking their breakfast of fruit and young shoots. There are two or three varieties of the species, the commonest being the silver-grey kind, called by the Malays *Wa-Wa,* on account of its note. There is also an inland variety, with the hands and fore-part of the body black or dark-grey, and with grey sooty markings which in certain lights give them a uniform dark-ash-coloured appearance all over the body, and from which they get their local name meaning "coal coloured." Those found nearer the sea are brown rather than grey, in some cases the abdomen and chest being of a light yellowish hue.

Gibbons make excellent pets, being easily tamed and rarely vicious; even when they are long accustomed to living in houses, they preserve their arboreal habits, and swing from rafter to rafter of the native Long Houses with an ease that makes the human acrobat look like a Robot. But this is not the only point about these pleasing apes which marks their kinship with mankind; for they, like us,

A GIBBON (Hylobates muelleri)

PEN-TAILED SHREW
(Ptilocercus lowei)

have what is called stereoscopic vision—in other words, their eyes are so placed that they can look directly in front of them, while a horse or dog (for instance) can only look sideways. Besides this, like ourselves, they can distinguish different shapes and colours, much more readily than other animals; born and bred in the primeval forest as they are, they have necessarily to be quick to spot any sudden movement or new feature, which to them may mean life or death; for the same reason their powers of hearing are almost miraculous.

The most remarkable feature of the apes is, however, the hand, which is very long in comparison to its width, while the thumb is very short. The fingers are flat, but have a deep crease running across the palm, while all the fingers are adapted for gripping purposes, and the Gibbon can do little more than bend and unbend its hand; but this power is of great use to it in swinging from branch to branch. In addition the apes have a highly developed sense of touch, so that their eye and hand work together even more than those of a squirrel.

At this time of the morning one frequently sees numbers of Macaque monkeys coming to drink at the many streams. Of these the most common is the pig-tailed macaque, called by Malays by the name of *Brok*. The general colour is a pronounced olive, in some cases shading off into brown, the altered tinge being perhaps caused by the development of the yellow and black rings on the hair. The upper portion of the head, and the line on the back and at the base of the tail are deep brown, while the face is a dusky flesh colour. These monkeys are most intelligent little creatures and are easily trained by the natives to collect coconuts. They are sent up to the tops of the fruit-bearing trees with a long cord tied round their waists; a pull on this cord by his owner, coupled with a peculiar grunt or cry, indicates the ripe nuts, which the monkey promptly picks or twists off. After a time they get so proficient as to be able to dispense with the cord, trusting solely to verbal orders.

On the other hand, in a wild state, the male macaques are often exceedingly savage and when confronted with a pack of dogs will stand at bay against a tree and severely wound or even kill the oncoming dogs. When seeking new haunts they go about in droves, the big males leading; occasionally one may see a solitary individual (like a rogue elephant) who has been ousted from his chieftancy by a younger rival.

Another commonly met member of this group is the crab-eating Macaque (*Macacus cynomologus*), which is found widely distributed over the whole of Malaya, both Peninsula and Archipelago, and also in Lower Burma and Siam. Its diet consists of fruit and insects, and, along the coast, of the small crabs found in the brackish water of estuaries. Unlike most monkeys and nearly all the apes, it is a particularly adroit swimmer and diver.

Perhaps, however, the most interesting of all is the Lion-tailed Macaque (known locally as *Gnumbo*), so called because his short tail is tufted at the end like a lion's. In main colour he is usually black or blackish-brown, with an enormous grey mane surrounding the face; in the young there is no mane, and the face is light flesh-coloured, and the baby when in the arms of its own parent looks like a changeling. The adult has powerful teeth and not only does a deal of damage to various crops, but is difficult to capture and is said to have a sulky temper when domesticated. He appears to be merely a large variety of the Pig-tailed monkey, having most of the characteristics of both the larger and smaller varieties of this species nemestrinus, of which the Brok is an example. He is a fine upstanding creature, some of the old males weighing over twenty pounds compared with the five or six pounds which is the normal weight of other members of the species.

A very handsome family of monkeys is that of the Langurs, the best known of which is *Semnopithecus entellus,* the "Sacred Monkey" of India, so called because these monkeys are said to be those who, in the Ramayana, helped Hanuman to build the bridge for Rama between India and Ceylon. The name of the group is *Semnopithecidae,* and they are found in any part of Asia where there are extensive forests. In Borneo there are at least seven distinct species, including a very fine black and grey variety named after myself, and first found by me. It is a handsome largish monkey, the general colour being a hoary grey, the hairs being mingled black and white; the hands, feet, crown of the head and face are a deep black; the rest of the head, *i.e.*

the forehead, temples and front of the neck, and also the inside of the limbs to half-way up, are pure white. The three contrasted colours are very striking. Another species, Everett's Langur, is very similar except that the white is everywhere replaced by yellow. Curiously, this animal is only found in the higher mountains, above 3,000 feet, but Hose's Langur is often seen in the low country, to which it descends at certain times in search of fruit.

A third variety is *Semnopithecus cruciger,* a beautiful creature, the first example of which I found at Bakam on the coast of the Baram district. It gets its name from a black cross on the back of the neck, back, and arms. Its colour is a deep brownish-red, with whitish fur on the lower parts of the body. The cross pattern develops late in life, for many of the young ones are of a pure gold hue, and some of a pale yellow, almost white.

Closely allied to the langurs is the Proboscis Monkey (*Nasalis larvatus*) a grotesque long-nosed creature found in the forests of Western Borneo, and possibly on the Eastern side of the island; so far as is known at present it is not found anywhere else in the world. They are, however, fairly common along the west bank of the Sarawak River, where they may be found sitting half-hidden among the leaves and branches of high trees overhanging the water, so closely blending with their background that they are very difficult to obtain. The general colour of the Proboscis Monkey is chestnut and yellow-brown, while its long tail and the lower parts of the body are whitish. Its most prominent feature, however, and one which makes it, after the Orang-utan the most striking creature of the Malay Archipelago, is its curiously developed long nose, on which account the natives call it *Orang Blanda* (Dutchman), though I am not aware that the Dutch are the exclusive owners of such a feature. This appendage, which is almost a caricature of a nose, is only fully developed in the males of the species; among the females it is shorter, while in the young it is squat and turned up, almost like a pig's snout. Early accounts given of these monkeys represent them as jumping from tree to tree (which they certainly do with great agility) and holding their noses to guard them from injury. This may very well be the case, but all I can say is that I have never seen it myself.

Up to this point, we had been traversing country more or less frequented by man, and at least partially inhabited, but now, as we left the river and approached the foot-hills which surround the base of the mountain, we had to cut our way through thick undergrowth and a perfect lattice-work of creepers. It was only occasionally through the dense foliage above us that we could catch a glimpse of sunlight. Although we knew that everywhere round us the forest was teeming with life beyond man's computation, all that we were immediately aware of was gloom, silence and an oppressive emptiness. In such circumstances the amateur, if he wishes to find bird, beast, or even insect, must perforce look for it. Yet to the eye trained to the jungle, there is abundance. Hanging from the branches of the gigantic Tapang tree, which from a base perhaps fifteen yards round rises to a magnificent trunk over two hundred feet in height, may be seen bees' nests; these are the homes of the wild bees (*Apis dorsata maniye*), a fairly large insect, and *Apis nigrocinata*, called by the natives *Nuang*.

This latter makes a very small comb but produces a great quantity of honey, which though not of a high quality is much relished by the Punans. It is a pretty sight to observe them clambering up the great trunks on scaling-ladders of bamboo attached to hardwood pegs driven into the trunk. When the objective has been gained and the nest secured, the successful prospector takes a delight in letting down chunks of the honey-comb to his friends below, at the same time indulging in a pantomime indicative of the pleasure which he gains from the honey and (what is estemed an even greater delicacy) the grubs.

Other insects which are fairly easy to discover, are the various kinds of ants, with which Borneo seems to be especially favoured. Among the worst are the so-called Fire Ants, whose sting is like a burn, and who will be found crossing one's path in literal thousands. I have often stood watching them and wondering where they came from and where they were going; without a break in the procession, they would march down the trunk of some great forest tree, across such path as there was, up a bank and then up another tree, an organised body heedless of other's concerns. A sort of first-cousin is the large red ant, known to Borneans as *Kessa,* which builds its nest in the higher branches of small trees, binding together the living bunches of leaves into a sort of envelope. These nests prove a great attraction to the local tribesmen, who consider the larva as a great delicacy; as soon as they spy one of the nests, they shin up the tree, heedless of stings, and chop off the branch on which it is built. Meanwhile their friends below have kindled a fire, in which they singe and burn off the ants, then, breaking open the bundles of leaves containing the nest, regale themselves with the fat young grubs, like an oyster-lover on the 1st of September.

Black ants of all sizes are common. One kind, known as the Elephant Ant, a giant about an inch long, with a large head and strong jaws like pincers, will run over your arms and hands without attempting to hurt you; but directly you touch them or make them think you are going to harm them, they bring their nippers into play; and then you know all about it. Another kind, not so virulent, but equally nervous, will stop suddenly if you as much as shake a leaf of the rattan or other creeper on which they are travelling; suddenly you hear a whirr as if all the factories in Burnley had suddenly gone mad. I imagine the noise to be one of warning; and I have certainly always taken the hint and acted on it.

Termites, or White Ants so-called (which really are not ants at all), are to be found everywhere. There are, in Borneo alone, many varying species of these fascinating but destructive creatures, each sort constructing a somewhat different type of nest. Some build on the ground and raise a yellowish-black mound two or three feet high; others construct black-coloured nests on the ground, or in trees, while others make nests shaped rather like the old-fashioned soda-water bottle.

Termites feed on various kinds of vegetable matter, fungi, wood, and (when obtainable) official papers; they seem particularly to love damp, warm, sheltered spots, and would be quite unable to withstand a prolonged winter's frost. Their chief enemies are bears, birds, small mammals and ants; their weapons and defensive powers are almost non-existent, but they realise perfectly that there are two kinds of protection—one the warding off of

TERMITE QUEEN, KING, WORKER AND SOLDIER.

TERMITES' NEST

attacks, and the other the more important, their prevention. It is on the latter that they chiefly rely; hence their complicated defence works.

While the various kinds of termite have each their own differentia and variations, every species has four classes or "castes" of insect, living together and playing their part in the community. The first of these are the Males—the term is a rather awkward one because the creatures are by nature unfitted to become fathers, being "strictly neutral," although generally they live with the females. They have one pair of compound eyes (like a fly), and when young have two pairs of large almost membranous wings; these are shed in later life, leaving only the stumps. They are also commonly dropped in rainy weather, when one's bungalow lamps prove a fatal attraction. At such times they become such a nuisance that the only refuge from them is to put one's lamp outdoors, and near it a large bath-tub. By such means, anywhere in tropical Asia, one may trap hundreds in as many minutes.

The female is very similar to the male, except for size. But besides these two sexes, there are two other "castes," the "Workers" and the "Soldiers." Neither have wings, and the soldiers are usually sterile. Their business is defence. Against ants, the worst enemy, they protect the home by means of a thick heavy fluid which they secrete in a sac in the head; against larger creatures they can do no more than offer their bodies as a living rampart, filling the spaces between the fortifications which the workers are erecting. M. Maeterlinck, in his admirable book, has given a vivid picture of the "soldiers" defending the outposts against attack, while the "workers" (who, by the way, are for the most part blind) simultaneously erect an inner wall or fortification, leaving the "soldiers" to die at their posts.

I once came across a body of termites engaged in biting off small pieces of green leaves and carrying them back to their home in the earth-mound. Suddenly it began to rain; the Union promptly "downed tools" and retired to their homes and to funds already accumulated; when the sun shone out again, they resumed their labours, not merely collecting arrears of work, but carefully rejecting such fragments as were too wet to be useful.

The centre of the Termites' nest is the "royal" cell, to which, apparently, all these labyrinthine and laboriously constructed passages lead. Here live the King and Queen, as the "perfect" male and female are called. These creatures when first emerging from the pupa stage, are winged, but after a short flight lose their powers of original and independent movement, and for the rest of their lives live prisoners in the Vatican built about them by their supporters.

At the period of maturity, the body of the Queen swells to a prodigious size, until she becomes a great, white, cylindrical creature, in shape rather like a sausage, and vaguely reminding one of the House of Commons when three-line whips have been issued. Here, however, the resemblance ceases, for the Ant-Queen is not merely a talker. She lays eggs at the rate of about ten thousand a day, a lesson in Mass Production; the eggs are carried off by a Child-Welfare section of the workers, whose task it is to care for the helpless young. To build up the Race these Workers (whose name is Legion) at once proceed in the true Imperialistic Spirit to enlarge the nest, constructing along the trunks and branches of trees tunnels of clay through which they bring to the crèches material such as gum, leaves, and decaying wood, for the coming generation. In their journeyings to and fro, each of the Workers is accompanied by a Soldier, like a plumber's mate.

Curiously, it may be observed that in the midst of this democratic society are found various kinds of grubs and worms of an alien nature and habit, living apparently not merely in harmony with the rest of the community, but to the mutual satisfaction of each. The Termite is, in fact, as Aristotle said of man, a naturally social creature; but since his value to the world is at least problematic, one may doubt his usefulness as a political guide.

Our way through the forest was cut in a sort of twilight, and we rarely saw signs of either Beast or Man, for the inhabitants of the jungle are very shy, and if they hear footsteps they vanish before you can see them. Sometimes, however, there might be the track of a wild animal, and, more rarely, some sign, such as a bamboo slit in a particular way, a sort of sign-post put up by a forest-dweller; I should not have remarked them myself, but my followers were trained to observe and ready to point out all such clues. Birds we could hear high above us in the gigantic trees, but we rarely saw them, except occasional Sun-birds, flitting among the underwoods, twittering and coquetting with the flowers as if to coax them to yield up their honey, or perhaps a lurking insect. For this their long slenderly curved bills are admirably fitted, and they are of equal use in the building of the curious nests which this bird constructs. A nest appears at first sight to be composed of skeleton leaves, attached to the under-side of some large leaf; but this is not the real nest, but only a covering. Both it and the home proper are slung on to the leaf by silken threads, stolen from spiders' nests, and fastened by the bird's bill, which it uses as a needle or bodkin. At one end the covering is bound close and tight, but at the other the threads only hold it together loosely, so that the mother-bird can creep into the nest safe from observation and out of the way of snakes, which are her greatest enemies. It is an interesting fact that this bird in finishing off the ends of thread, uses a "Malay knot," of the same kind as all Malays make. The ordinary man if he wishes to secure the end of a string and prevent it from slipping, makes a knot by passing the cord's end round itself and then through the bight thus formed, but the Sun-bird coils the end of the thread round and round on itself, rather like a Chelsea bun, or a worm-cast, and just as one screws a piece of paper into a pellet. The method sounds somewhat crude, but it is quite effective; I imagine that the Malays copied it from the bird, although, when one considers the imitative power of sea-lions and apes, one should be surprised at nothing.

An interesting relative of the sun-birds is the Racquet-tailed Drongo, a blackish-blue bird of a larger size, known elsewhere and more generally as the King Crow, although he is, of course, not a crow at all. For his size he is perhaps the cheekiest bird that one knows, the common starling being among the "also ran." He is seen at his best in driving away larger birds, especially birds of prey; for he possesses neither shame nor fear, and by sheer bluff he puts everyone off his stroke. I am sure he would make a good pitcher at baseball. The Drongo is often called the servant of Kra (the little macaque monkey) from its habit of following a troop of these monkeys. It does not accompany them out of altruism, but in order to catch the various small insects which the monkeys disturb as they jump from tree to tree. Another interesting fact about this little bird is that it is mimicked by a bird of the cuckoo family, which deposits its egg in the nest of the Drongo. The cuckoo's purpose seems to be a double one—the first to farm out its nestlings and get them free board and lodging; the second to protect them against hawks and other birds of prey, for the Drongos are not the least afraid of these birds, but will attack them in mass-formation and drive them from the field.

I should think that there must be at least ten species of Cuckoo indigenous to the Eastern islands, but none of them are in the least like the well-known European bird, either in their habits or in their call. It is interesting, however, to note that the European cuckoo during July and August migrates to Eastern Europe and India, and, on rare occasions, has been found as far afield as the Malay Peninsula, and even Borneo. The cuckoo of the Peninsula neither lays eggs in other birds' nests, nor is his call in the least like theirs. Two species that I know build nests for themselves, and lay eggs varying in shade from bluish-white to pure white.

A Cuckoo common in Borneo is known to the natives as *Kapa Kapong,* whose activities force themselves on one's notice at the time of year when certain flowering trees blossom. As soon as the buds burst, this bird is found picking up insects from the flowers, and every time it catches an insect it emits a sort of deep note, *Kong-Kaput,* or *Kapa-Kapong.* When several of them gather together and indulge in Community Singing (which they generally manage to do as near one's bungalow as possible) they nearly drive one mad, especially if one happens to be recovering from an attack of malaria. Even worse, however, is another cuckoo, which I am told is known in India as the Brain-fever Bird. His repertoire consists of a cadence of three notes, which he repeats, gradually going up the scale. You listen, wondering when he will reach his limit; but, just when you think he is about to give up, he drops down the scale and begins all over again.

A cuckoo with a solemn note like a bell is known as the *Mindu.* He is a medium-sized bird with a beautiful plumage of iridescent rainbow-blue, shot with a tinge of grey; his head bears a small patch of tiny bright red feathers.

But the most curious call of all these birds is that of the *Enterakup,* who lives on the coastal plains and has a note like *Kup-kup-kup,* from which it undoubtedly gets its name.

Another fairly common cuckoo is the *Bubut,* better known as the Crow Pheasant, a large bird with a brownish body and a black head, about whom and the Argus Pheasant the natives have a legend. The two birds agreed to tattoo one another, but the Argus Pheasant, having been magnificently tattooed by the *Bubut,* left him in a bath of tan made from mangrove bark and refused to carry out his contract, being alarmed by outside enemies. The sob-like cry from which the Bubut gets its name is a lamentation for its dingy plumage.

The Argus Pheasant comes into another story with another cuckoo, a beautiful and rather rare bird which lives chiefly on the ground, and has a gorgeous purple-blue plumage on its back, while the chest feathers are barred with stripes of grey and white; its legs are of a bright jade-green. This bird is known to the natives as *Kruai Manang,* which means the Doctor of the Argus Pheasant; for he is said to have removed the curse of sickness which befell the Argus Pheasant after his scurvy treatment of the *Bubut.* *Kruai Manang* holds a high position in the Bird Aristocracy, according to legend; and, by his beauty he certainly deserves it.

CHAPTER VI

A MOUNTAIN TREASURE HOUSE

Salt-licks—The Bear-cat—A Reconnaissance—An Edible Monkey—Tree-shrews—Offensive Protection—The Leopard-cat—The Barking Deer—Broadbills—Hornbills—Their Nesting Habits—Providing for the Widow—The Rajah's Breakfast—Spiders Bigger than Birds—A Curious Orchid—The Uses of the Woodpecker—The Smallest Owl in the World—An Invisible Target—Gossamer—The Storm Spirits—The Peregrine Falcon—Collector Turned Naturalist—Flying Squirrels—Some Rare Cats—The Mountain Babbler—Bald Birds—Francolins—Overlooking the Forest.

DURING the course of this day's journey we were agreeably surprised to find a sort of ready-made path cleared, as far as one could guess, for our special benefit; on either side of the track the bushes were sprinkled with mud. On making enquiries I was told that a rhinoceros or some other large creature had passed that way. The Bornean rhinoceros is a smallish species and quite the most grotesque of his kind; he has two horns and his hair is tough and bristly, almost like fine wire. He frequents the foot-hills below the mountains, and comes down in the heat of the day to take his ease in what are called " salt-licks," muddy baths formed by springs of saltish water. The clearing and the mud on the bushes were, I was told, caused by the creature's trampling movements on his way home to his lair higher up the hills. My informants were quite right, for very soon we came across one of these salt-licks, although we found no one " taking the course." However, the Langur named after me (*Semnopithecus hosei*) frequents these spots, partly for the water and mud, and partly for the saltish taste. A peculiarity about this agile little monkey is that in its gall-bladder are found the curious Bezoar stones—hard, brittle, oval-shaped concretions of a dark olive-green colour, and much prized by the Chinese for their medicinal properties; there is probably some chemical substance in these " salt licks " which is said to help in their formation. A soft form of stone, but otherwise similar, is found in the intestines of the Bornean porcupine; while another substance rather like Bezoar stones, and also much valued by Chinese, is found in the bodies of animals (usually monkeys) which have been shot, but not immediately killed, by poisoned darts from the blow-pipe; very often, in the centre of the bezoar a portion of the dart, sometimes an inch or more in length, may be found encrusted in the materials.

This same day we came across a fine Bear-cat (*Arctitis binturong*) swinging itself about in the branches of a tree with the help of its prehensile tail; apparently it was in quest of food, a rather rare thing in the daytime as the animal is usually nocturnal. This quaint cat-like animal is black in general coloration with a grizzled head, it has an extremely long tail, longer than its head and body together, covered with straggling hairs longer than those on the body. It usually lives in the trees of the forest, on a diet of small mammals, birds, earthworms, insects and fruit; of the latter they are very fond, and in search of it will leave their ordinary haunts and raid gardens and orchards for bananas. It is easily tamed and is often kept as a pet by the natives; when it wants to be fed it will sit up on its hind-legs, flapping its paws and begging like a dog.

It will be understood that in such thick jungle as we encountered progress was slow; and it was, in fact, not until the fifth day of our overland journey that we arrived at the foot-hills. Here we decided to make a sort of semi-permanent camp or base, so as to have somewhere to keep our stores. Almost all the night before our carriers had been kept busy bringing up the heavy bags of rice; now they had a rest while the remainder of us started erecting our depot, the framework being of small tree trunks, and the roofing leaf mats, which if well-made are almost waterproof. The spot we chose for our camp was well selected, for there was a beautiful waterfall near by which ensured a good supply of clean, clear water; and I suggested that we should spend two nights there before we began the actual ascent. It was advisable, obviously, to make some sort of preliminary survey before we started climbing; but what was more necessary was that we should dry our clothes, which were absolutely drenched. Cutting down a few of the smaller trees in the immediate neighbourhood, we cleared enough space to enable the sun's rays to pierce through for a few hours each day; and the next morning, while the rest of the party were putting up our huts, I took my field-glasses and made a reconnaissance.

I was pleased to find that the mountain was a sandstone one with a sloping side, for sandstone hills are generally much easier than those of limestone, of which there are many in Borneo. When we actually started, however, on the next day, we found the going fairly easy but very tiring, for our route lay over a succession of small hills, rising to a height of about a thousand feet; and after several hours of hard foot-slogging I found from my aneroid that we had not reached an elevation much higher than that from which we started.

All around the scenery and vegetation were gorgeous; the forest was still dense, but we encountered a succession of mountain streams with beautiful waterfalls. Towards evening, before halting for the night, we came upon a small party of Punans who had been out hunting with their blow-pipes and had obtained two specimens of that pretty little grey monkey, *Semnopithecus hosei*. The Punans eat the flesh of this and other monkeys, which they roast whole after burning off the hair. Knowing of this method of cooking I begged for the skins, which they were pleased to give me. In a very few minutes I had the skins carefully taken off and the skull and bones of the leg and arm removed; the rest of the carcass was then returned to the hunters, who very quickly stuck skewers into the flesh and roasted it over the fire.

On the next day of our journey almost one of the first creatures that we met was a tree-shrew (*Tupaia montana*). In colour it is a dusty olive, with a strong rufous suffusion; the head is of a clearer olive hue and has a deep black line running down the back. The under surface of the body is of greyish orange. The mountain species differs from the others by the colour of its tail, which is shorter and of a bright chestnut colour. Another remarkable tree-shrew which I found near here was *Tupaia tana*. The fur of this little animal is long and fine, and he has a " double coat " like a Scottish terrier. The hair is of two kinds, part long, and quite black, the rest shorter and with a band of orange or dark red-brown near the tip. One of the most interesting and rarest members of this family is the Pen-tailed Shrew (*Ptilocercus lowi*), a tiny grey creature with a very long, slender tail, quite bare except for a tuft at the end which gives the appendage the appearance of a quill-pen. It is named after Sir Hugh Low, in whose house on Labuan Island the first specimen was caught. It is, in fact, usually in and near human habitations that the very few known examples have been found. I myself was lucky to obtain one, which came into our hut when I was camping somewhere at the back of nowhere, and which was struck by one of my men and captured. This shrew, like many others, appeared to be distasteful to other animals and birds, for one of the few specimens known was caught by a dog, who immediately dropped it in disgust.

The sense of disgust, however, whether of appearance, smell or taste, reaches its zenith in the case of *Gymnura rafflesii,* the stinking shrew, a nocturnal creature who is a sort of shrew but about the size of a rabbit, "resembling," says Mr Shelford, "a big white rat with a long pointed snout." The body is clothed with a scanty white fur, but the tail is nearly naked; some varieties are black. He is one of the best examples of what one may call Offensive Protection. He possesses no armour, whether for defence or attack; but his appearance, his loathsome acrid smell, and his colour, are warnings to possible foes. He positively flaunts his unattractive qualities, and enjoys a proportionate immunity.

Among the animals which we met here were the mouse-deer and *Hemigale hosei.* The latter is a pleasant little animal belonging to the civet-cat tribe. Its colour is a uniform smoky brown or black, the bases of the body hairs being whitish and the long whiskers white. Its chin is white and its ears, which are thickly covered with hair, and the rest of the limbs are black. Here also we found *Felis bengalensis* (the leopard cat), a small animal about the size of our domestic cat with a smooth coat, in colour varying from reddish-grey to brown. The natives keep a sharp look-out for it, for it is very fond of stealing fowls, and in the lower country is often found prowling about under the houses. When captured, it is very shy and nervous, crouching and swearing like a wild cat; but a tame specimen is known to have lived in London for six years.

This day my friends shot a fine specimen of Kijang, of the same family of deer as the Muntjac or Barking Deer, but in colour not at all like it, but more like our own common deer.

SCOPS BROOKEI

HEMIGALE HOSEI

I was struck, in this higher region, at the beauty of the bird-life. There were numbers of busy flitting little barbets, exhibiting in their plumage the most violent and striking contrasts, red, blue, purple, and yellow on a green ground, generally with coloured cheeks and a patch of bare skin round the eyes. Then there were various kinds of Broad-bills (*Euryloemidae*); these birds have an extremely wide range, being found on the lower spurs of the Himalayas, and also throughout Burma and Siam. Distinguishable by their bright colouring and the broad beaks which give them their name, they inhabit the jungle in the neighbourhood of water, feeding on fruit, and also worms and insects, the latter being often caught on the wing. Their nest is a large untidy structure, but most ingeniously constructed, and is often found suspended from a low branch or a bush overhanging a stream. In shape it is oval, with an opening near the top, and is often with an overhanging roof as protection; it is built of twigs, roots, and leaves and is lined with finer material. Four or five eggs are usually laid; sometimes six, spotted all over with light or creamy white markings. When I first made the ascent of

Mt. Dulit I was fortunate enough to discover a new species of broadbill (*Calyptomena hosei*), and also its nest, which was hanging from the end of a branch about thirty feet from the ground. The bird itself was green with black spots on the wing-covers. The breast feathers were of a bright blue almost of a Cobalt shade; the under part of the wings was of this same gorgeous colour, which of course was only visible when the bird spread its wings. Otherwise its bright green colouring makes it inconspicuous among the thick foliage of the tall trees.

The most interesting birds, however, of this locality, and indeed one of the most curious family of birds in the world, are the hornbills. There is one species (*Rhinophlox vigil*) which has a solid beak about five inches long; it is found generally on the forest-clad slopes of mountains, where it can be heard calling with its peculiar note, *Te-gok-te-gok,* beginning deep down and gradually becoming shriller until it ends in a cackle of laughter. This particular bird has two long feathers in its tail which are greatly sought after by the natives, who wear them as an ornament only after they have taken a head in war. Ear-ornaments also are made from its beak, and like the feathers can only be worn after a head has been taken. The Punans have a legend that on the tree-trunk that bridges the chasm between Life and Death one of these birds is stationed. Those who in life have taken a head he helps over the bridge; but those who have not done so fall into the abyss and are miserably devoured by a giant fish called *Pitan*.

This hornbill is the only species whose beak is solid; in the other cases the casque is hollow or partially filled with a kind of bony tissue. In the case of *Buceros rhinoceros,* the casque is brightly coloured in shades of orange and red. One kind seems to preserve the colouring of its neck-feathers by means of a saffron-coloured oil secreted in a gland above the base of its tail, in what in a " table bird " is called the " Pope's nose," and transferred by rubbing with the beak.

The nesting habits of the Hornbills are most interesting. All the species build their home, for protection, in a hollow tree, communication with the outer world being by means of a slit or hole. If this opening is not at the right elevation above the floor of the nest the birds fill up the interior with leaves and twigs until it is of the right height for the mother-bird to be able to sit comfortably on her eggs with her beak protruding so as to receive food. This having been arranged satisfactorily, the hen bird spreads a thin layer of feathers plucked from her own body on the built up floor, and is then completely walled up by the male, who plasters over the opening with a sort of gummy substance which he secrets in his stomach; this substance hardens on exposure to the air and shuts in the female until her beak alone shows. In this uncomfortable position she remains a prisoner until her nestlings are from two to three weeks old, the male feeding her meanwhile with insects, fruits, seeds, and parts of frogs and lizards, all rolled up into a sort of pellet, which he throws into the expectant beak of his mate.

When feeding the female, the cock bird clings to the bark of the tree, or else perches on a convenient branch, and jerks the food into his wife's beak. I knew one instance where, the husband-bird having been shot by hunters, other males came and supported the widow. While the feeding and imprisonment process is going on, several seeds naturally are not caught by the hen, and falling to

THE BORNEAN RHINOCEROS HORNBILL. *Buceros rhinoceros.*
See page 28, *chap. vi.*)

the ground germinate; by observing the growth of these, the natives can infer the age of the young birds without seeing them.

When the time comes for the young birds to leave the nest, the mother-bird breaks away the plaster with her powerful beak, after which the opening is again sealed and both parents devote themselves to the feeding of the family. While the young are helpless, however, the sealing up of the mother is an excellent protection against snakes, monkeys, and many tree-haunting carnivores. A pair of Hornbills will use the same tree and nest year after year, which is not to be wondered at when one considers the difficulty of finding a suitable site and the trouble which has already been taken in building the nest. I should add that, as far as I know, no one has ever yet found a hornbill in the act of building a new nest, or of excavating a tree-trunk for the purpose; I myself believe that they do the same as woodpeckers, and, having found a tree which is partly rotten, assist nature by digging it out until they have provided themselves with a suitable site.

Hornbills' eggs vary slightly, according to species; those of *Buceros rhinoceros* are of a pepper-

and-salt appearance, white mottled with brown, while those of *Anthracoceros malayanus* are pure white. The nestlings are hideous, naked, fat-bellied and awkward, but are much in request among the Dayaks, who eat them raw.

The native method of catching the mother-bird during the period of incubation is rather brutal. The tree is scaled and the entrance broken open; the frightened bird flutters up the hollow trunk but is brought down with a thorny stick which is thrust in after her and twisted about until a firm grip is obtained of both her flesh and plumage.

Hornbills make amusing and interesting pets, for they become quite tame and will follow their owner about like a dog. One that I had at Claudetown became quite an expert at catching bananas thrown him. On one occasion the late Rajah was staying with me and was surprised over his morning tea and fruit by my pet, who coolly perched on the railing of the verandah. Taken aback by the grotesque creature, he looked round for something to throw at it, but while his head was turned for a second, a whirring noise was heard and his breakfast was gone.

This evening, while wandering about with no very definite purpose, I came across the large black bird-eating spider. I was sitting by the side of a small stream watching a pair of otters playing in the water and diving and darting with inimitable grace, when I noticed the trunk of a large tree which had fallen by the bank. Its roots not being deeply set, had broken off short but remained bound together in a flattish mass standing two or three feet above the ground and covered with a net-work of plants and ferns. Suddenly out of a hole in the bank where the tree had fallen I saw a large black spider with hairy legs appear and attempt to capture a tiny sun-bird which was hovering in front of a trumpet-shaped flower. The contest was rather interesting to watch, for the spider was slightly larger than the bird. As I watched, the spider moved warily along towards the bird, evidently trying to disguise itself, but the little bird smelt danger and called on a friend; and the two together started to tease the spider, which stood there quite still and with its legs held quite close to its body. The two parties gradually edged closer to each other, until suddenly the spider opened its legs and clutched at the nearest of the two birds. The little creature was however too wily, and escaped; whereupon the spider, probably highly peeved, disappeared into his hole, over which he had begun to spin a web.

Approaching nearer I began to investigate, and discovered four or five other holes, which presumably were other doors of the same house. As two of these holes had no web over them, it seemed clear that the spider had obtained enough for his immediate needs and was probably enjoying his meal at his ease inside. I now started digging at the hole where I had seen the spider go in, first of all poking down a long thin twig into it to find out which way the hole ran and also the depth of the hole. Leaving it in the hole as a guide I dug down as far as I could, and then again pushed it in to see how far it would go, turning corners as I proceeded. At last, at about four feet from the entrance, I came on the spider and was able to capture him without getting bitten. It now struck me that there might be more than one spider in the nest, and, one of my men happening to come up at that moment with a muzzle-loader and ammunition, I got him to give me a pinch of powder, which I placed in one of the

(Upper) FISHING FOR GREY MULLET
(Lower) A CURIOUS ORCHID RESEMBLING A SWARM OF BEES

(See chapter VI.)

holes where the web had not been disturbed, and setting light to the powder, jammed a clod of earth over the hole. When, later on, we dug out the hole, we found a second spider suffocated.

We camped this night at an elevation of about 2,500 feet in a pleasant spot with some curious orchids; one of these was almost exactly like a swarm of bees; another was like a spider, and a third mimicked the stick insect, which in turn mimics a plant. I can suggest no reason for these curious resemblances.

When I woke next morning, one of the first sounds that reached my ear was that of the woodpecker, the hammering of his beak on the tree trunk or bough being followed by the self-satisfied laugh of the bird as it flew from tree to tree in its unremitting search for insects. In his hunt for food he is the forester's most efficient ally in the destruction of bark-beetles and other insect pests, and wherever he is found, or wherever forest timber is valued, he should be accorded the protection he so well earns. Even in England, where the bird is well enough known, his services are not adequately rewarded; in the great timber forests of the world the amount of good he does is inestimable.

The woodpecker is equipped with a beak strong enough to cut into all but the very hardest wood; during his drilling operations his strong feet cling securely to the branch or trunk, while the stiff quills of his tail maintain his balance. The bill, by its shape, suggests a chisel and with it the bird delivers repeated vigorous blows that make the chips fly out in all directions until the hiding-place of the larva is exposed and the insect drawn out and swallowed. In addition he possesses a long, slender tongue, the upper surface of which is furnished with small backward pointing barbs, while the tip is as hard as a bone and as sharp as a spear-point. With these formidable weapons, the rapidity with which insects can be extracted and destroyed can be easily understood. Bornean woodpeckers have no fear of man and will go on hammering away in his presence quite without concern, considering him, doubtless, an intruder in the forest.

In Borneo there are more than fifteen species of woodpecker, varying in size and beauty from a large ashy-coloured bird about the size of a rook, to the tiny little omen-bird known as *Sasia abnormis,* which is not much bigger than a jenny-wren. This little bird is known to the Ibans as *Katupang* by reason of the shortness of its tail; it is considered one of the most important of the omen-birds. Like all other woodpeckers it nests in the holes of trees, laying long-shaped, white eggs to the number of five or six, as may be. Other fine species of woodpecker are one in which the male has gorgeous golden-yellow feathers on the back near the rump, those of the female being white; another, a large black and white bird with beautiful red markings, very much like the Great Spotted Woodpecker in England; and others with brown, yellow and red, and olive-green and speckled livery. One tiny variety, known as *Blatok,* has a darkish speckled plumage of black and white. There is also one kind known as *Kotok* which has two quite distinct calls. One of the notes, *Tok-tok-tok,* is considered a bad omen; the other, usually uttered when the sun is bright, and sounding like *Keing-keing,* is looked on as auspicious.

It is a curious fact that, according to Dr Warde Fowler, the ancient Romans (or more probably the

Etruscans) held the same views about the differing significance of the two notes.

At this point in our journey I was rejoiced to discover two birds, each of them the smallest of its kind known. The first of these was a liliputian owl known as *Glaucidium borneense,* about the size of one's thumb; and also a tiny hawk, *Microhierax,* which lays a large white egg about as big as itself. The entire forest at this point was covered with a thick matting of moss, most of the trees being either completely dead, or having only a few leaves at the top. Both these little birds settle on the dead trees; and as these are of a notable height, they look like insects, being in fact very much smaller than some of the large butterflies. One only observes them by noticing them when they settle on a branch, and then one aims at the spot where one has seen them settle, without actually seeing the object itself. (Of course, one uses dust shot with a considerable spread.) It is, however, rather strange to shoot something which one does not see.

The following morning we decided to move our camp up to about 5000 feet; our progress was slow as there was still a great deal of cutting needed, and also it was often hard to obtain a foothold. There was not much to see of animal life at this stage, but at one point we surprised an animal which looked like a barking deer, which stood staring at us for quite a minute before he moved away. I have since thought that it was a kind of goat, known locally at Serow. The most curious thing, however, that I noted was a young spider crossing a stream. We had halted by the side of a small mountain stream and I watched one of these workers paying out its silken gossamer threads until delicate life-lines several yards long were spun out and floating in the breeze. The spider seemed to stand upon its hind legs and when it calculated that a sufficient length had been paid out, it let its legs drop and was carried by a sort of natural parachute across the water to the other side, where the threads were caught on the branch of a shrub. In England on a sunny morning, especially in the autumn, one may find the same thing on a smaller scale, but in greater numbers, when hundreds of gossamer threads may be seen, spun by tiny spiders seized with a Wander-lust.

That day we obtained a specimen of the small

squirrel *Sciurus hosei,* a small mountain animal which moves up and down tree-trunks with short jerky jumps; it is a pretty striped creature, the body being of an olive-green grey, the hands and feet grizzled with orange and black, and the tail brightly ringed and suffused with black. We also caught sight of a pair of wild dogs, of a species exceedingly rare in Borneo, *Cyon rutilans;* they had been eating the remains of a young wild pig, but before we had time to get more than a glimpse of them they were away.

When we camped for the night and were waiting for supper, for want of anything better to do, I began to drum idly on the enamelled tin plates, noting the different sound of each. To my surprise, one of my men came up and begged me to desist, saying that I should draw down the anger of the Storm Spirit and that we should all come to disaster. As if to corroborate his words, about twenty minutes later a storm burst on us such as I have rarely seen. Gigantic trees and rocks came rolling down the mountain-side; all night long the lightning and thunder were incessant, while the rain added to our discomfort, and the consternation of my native followers, who, like all simple forest people, live in spiritual contact with the supernatural. We were all of us heartily relieved when day dawned and the storm ceased.

We were now very near to our goal and within a few hundred feet of the summit. Here it seemed that we were on ground untrodden by man, and occupied only by wild creatures engaged on their peaceful avocations. The peace is perhaps more apparent than real, for at about this height above sea-level is the haunt of falcons, especially the Peregrine, the most esteemed of the birds employed in falconry, the female being used for larger game, such as herons and rooks, and the male for partridges and smaller birds.

The Peregrine Falcon is an extremely rare bird in Borneo, and its existence in the island was for many years not known; it was, however, suspected, for about the year 1890 I received a letter from Dr Gurney of Norwich, one of the greatest authorities on birds of prey, prophesying the existence of an intermediate species of peregrine falcon, between the dark and the lighter variety, which he thought might be found nesting on the mountains of Borneo or Sumatra, at an altitude of about 4000 feet. Dr Gurney asked me if I thought it possible to obtain a specimen for the Norwich Museum, and suggested the regions near Mt. Dulit (5090 feet) as a likely habitat of the bird. A year or so later I ascended this mountain with my brother Ernest, and we obtained an adult bird and a young one of the intermediate species, which is now known as *Falco ernesti*.

The Peregrine Falcon has for me a peculiar interest, because it helped to change, almost completely, my attitude towards Nature. A year or so after the discovery of the bird I again ascended the mountain; near the place where I had found it I observed a rocky ledge, which I determined to explore, as it looked a likely spot for the Peregrine's nest. This I did by means of a rough ladder of rattan and rattan ropes fastened round my chest, by which my native helpers hauled me up. On reaching the flat shelf, tired and bruised, I was rewarded by finding what I had worked so hard for—a nest with four eggs.

I gazed on them with mixed feelings. At first the joy of discovery and the pride of the collector

PEREGRINE FALCON AND YOUNG

were uppermost. I had achieved my ambition and thought of the prestige I should win and the pleasure I should convey. I saw myself in the Bird-room at South Kensington, surrounded by enthusiastic friends gazing at these eggs, the first known to Science of this rare bird, and at myself, their discoverer. Then the emotion of the true Naturalist arose in me; and, between the two motives, I thought of a compromise, to take two of the eggs and leave two. But, I am pleased to think, the better feeling won; Conscience or Compunction asked me why this extremely rare bird should not be allowed to propagate its kind in peace, as Nature intended. I therefore merely noted down the appearance of the nest (it could hardly be called a nest, for it consisted only of a few sticks), the size and colour of the eggs, and then descended by the way I had come—no longer a Collector but a Naturalist.

While we were encamped here, the stillness of the night was constantly broken by the cry of the owl named after the late Rajah, *Scops brookei*. Its note is clear, and at first almost startling, but is repeated monotonously with no change of inflection. This is a rare bird, found only at considerable heights, and belongs to the family of Eared Owlets, for its ears are set back like those of a dog that has done something wrong; the position gives it a somewhat demure and deprecating appearance.

The Kayans have a bird-call, made from a short length of bamboo, which when blown is so exactly like the note of the real bird, that it will always answer back, and so to speak join in the conversation. A sort of note or key is made about the centre of the joint of the bamboo, which can be sounded by blowing down one end, with the hand covering the other end. About half-way through the sound of the note, the hand is released, giving exactly the finishing sound of the bird's call. It is quite ingenious, and the birds are deceived into the belief that another owl is present in the hut, for they will often fly inside, so realistic is the sound produced.

The next morning we started on the last stage of our upward journey, and soon we came across a pretty little squirrel, *Sciurus brookei,* named, like our friend of the night before, after Rajah Brooke. This pretty little animal, which is usually found at a height of about 5000 feet, seems to belong to the Oriental group of squirrels; its colour is a dull olive-grey, with no stripes on the sides of the body, but with black and yellow rings on the tail. Some have an ornamental reddish patch on the head, shoulders, hips and tail.

Borneo is, as may be imagined, remarkably rich in the matter of squirrels, in three species of which I have a peculiar and personal interest. These are *Petaurista thomasi, Petaurillus emiliae,* and *Sciuropterus hosei.* The first of these I found in the year 1897. It is one of the extraordinary flying animals found in Borneo, which I have described elsewhere. It is easily distinguishable from its congeners by its splendid ruddy colouring, which makes it one of the handsomest mammals in the island. As a compliment to the late Mr Oldfield Thomas, of the British Museum, to whose help and encouragement I owe more than I can say, and in whose room at South Kensington I wrote my "Mammals of Borneo" when I was on leave in 1892, I named the beautiful creature after him. Perhaps as a return, but in any case with characteristic delicacy, Mr Thomas named the other two squirrels after Mrs Hose and myself. Of these, *S. hosei,* also a "flyer," is very dark in colour, with the tips of the dorsal hairs washed with a reddish fawn tinge; its sides and the top side of the parachute are black. The under surface is white with a slightly rufous tinge. The feet are brown, but the ends of the toes are white. The tail is a dark brown which, as it approaches its tip, gets darker and darker, until at last when you would expect it to be black, it cheats you and changes abruptly to white.

P. emiliae, the smallest member of the group of flying squirrels, is similar to S. hosei, but very much smaller; its colouring on the back is somewhat lighter, and the lower parts are pure white. It is readily distinguishable by its size.

This morning we also found a fine Clouded Leopard (*Felis nebulosa*), a beautiful graceful creature which is found both in the low country and in the mountain jungle. Its general colour varies from a greyish or earthy brown to a lighter, more yellow shade, the lower parts and the inner sides of the limbs being white or pale tawny. The head is spotted on the dome; while, starting from between the ears, are two broad black bands with narrower bands and elongated spots between them, which run back to the shoulders, and are generally prolonged down the back. The long tail is densely furred and marked with a number of dusky rings. This animal is much sought after by the native tribes, the canine teeth being used by the Kayans and Kenyahs as ear-ornaments, while war-coats are made of the beautiful coat.

A perhaps even more beautiful member of the cat-tribe is the magnificent Marbled or Golden Cat (*Felis marmorata*). This animal has a very wide geographical range, being found not only in Borneo, but in the Himalayas and as far as Central China. It is a very beautifully marked creature, usually of a golden chestnut shade, with a pink nose and large topaz-coloured eyes; the cheeks are striped with white, the under parts of the body and the tail are pure white. Another beautiful variety of the cat tribe in Borneo is an animal known as *Felis badia,* who seems to be a red cat which has gone black—a case of melanism like the black panther. It is a fine gorgeously-coloured creature, a little bigger than the biggest domestic cat, and has a very long tail. It is an extremely rare animal, but there is one example at the Nat. Hist. Museum, presented by myself. Another rare cat is *Felis temminckii,* named after the Dutch naturalist; it is about the size of a large fox-terrier, and no doubt is to be found in Sarawak, although up to the present the only examples I have come across were from Dutch Borneo.

Another group of Wild Cats is that of the Civet Cats. In Borneo there is the Palm Civet, which is distinguished from other civets by the smaller teeth and the nakedness of the feet. Then there is the Malayan *Munsang,* which is distinguished by a light band across the head. The body of these civets is comparatively small and somewhat compressed, but the limbs are long in proportion. In colour they are grey with blackish-brown streaks and blotches.

The most curious of the species is the Water Civet, or *Bennet's Cynogale,* which is easily distinguished from the other members by the absence of a groove on the nose and upper lip. It is a peculiar little animal with beautiful fur, and of aquatic habits; but besides swimming, it can climb trees, and is said to thrive equally well on fish and on land animals or even bananas.

On this last day we came across two birds which presented an interesting problem. These were a male and female of the Mountain Babbler (*Allo-cotops calvus*), both of them quite bald-headed. Bald-headedness in birds is most uncommon and is usually difficult to account for. In the case of the Vulture and the Adjutant Bird it may be explained by the scavenger's life which the birds lead, but even so, in the case of the latter, one might suppose that the bird's long beak would suffice to keep the bird out of reach of the putrefying offal on which it feeds. But with the Babblers there seems to be no explanation; they have no feathers at all on the top of the head, not even scurf, but only bare dull yellow skin; and as the female is as bald as the male, there is no room for the theory of sexual adornment.

Perhaps the most unusual and most interesting of all these bald birds is *Pityriasis gymnocephala.* In its youth the bird is bald, but as it grows up it puts forth a few red feathers dotted about on the bare skin of the head. After a time these dis-appear, but in their place there come up small spines like those on a horse-chestnut, or like hair in the malady known as *Pityriasis,* a sort of scaly disease of the scalp; it is from this disease that the birds gets its name. The young birds have a livery of black and red on the chest and the lower parts of the body; but in the old birds, both male and female, the legs, thighs, and head alone are red, the rest of the body being an ashy black. At the same period hard black callosities like wattles are found on the cheeks and near the ears. These give the bird somewhat the appearance of a Mynah, for which reason (and also from the fact that it occasionally whistles rather like a Mynah) it is called by the natives "Tiong bali," which means "Phantom Mynah," "Tiong" being the name for the Mynah.

For a long time it was doubtful whether this bird belonged to the class of Shrikes or of Starlings; there is, however, an egg (which was found, unlaid, inside a specimen shot in Borneo) in the Sarawak Museum at Kuching. This egg is white, with brownish spots almost all over, but especially at the larger end. This is the only egg of the bird that has been found; and it would seem to indicate that the bird is a Shrike, corroborative evidence being found in the shape and size of the beak of the parent, and also from an unpleasant and brutal habit which it is said to have of turning upon its own friends and attacking them when wounded.*

Mr R. H. Burne of the Royal Coll. Surg. Museum reports that a section taken from a dried specimen shows the scalp to consist of a dense fibrous tissue containing one or two blood-vessels, but none of the appendages of the skin. The surface is covered with a layer of stratified epithelium which is teratinised throughout and in the superficial part consists of thin scales. Epithelial columns passing in from the surface to form papillae are absent. The small spines seen microscopically consist of branched and unbranched processes of the dense fibrous tissue of the scalp, and they are covered with a similar epithelial layer. The condition apparently arises from atrophy of the skin and its appendages. Whether the formation of the small processes is in some way connected with the atrophy, or whether they should be regarded as papillomata, is not certain; they show no signs of active growth, which is against their being considered as papillomata.

At about this height, *i.e.* about five hundred feet

* See frontispiece plate.

from the summit, I was lucky enough to shoot an interesting wood-partridge, known as the Dulit Long-billed Francolin (*Rhizothera dulitensis*), or alternatively Hose's Long-billed Francolin, so called from a pair which I obtained on Mt. Dulit at a height of about 4000 feet. It is a very rare bird indeed, for only the four specimens obtained by myself are known; this one flew out of a great bed of moss, which is apparently its habitat, and from which, owing to the humid atmosphere and the perennial darkness, it has acquired a very much richer colour than the commoner Long-billed Francolin known as *R. longirostris*. Its general colour is a chestnut shaded with black, but the breast is grey and the under-parts are white; its beak is long and curved and gives the impression of great strength. In contrast with the demure colour of its throat the head and cheeks are of a striking reddish chestnut, which give it a very noble appearance.

Other fine wood-partridges are the very rare *Caloperdix borneensis*, a fine chestnut and black fellow, each of whose body-feathers is marked with concentric white lines; and a bird with the strange name Roulroul (*Rollulus*), which is found fairly

YELLOW-BREASTED TROGAN
(Harpactes dulitensis)

HOSE'S LONG-BILLED FRANCOLIN
(Rhizothera dulitensis)

frequently in various parts of Burma and Malaya; its chief characteristics are a tuft of long hair-like bristles in the middle of the forehead (rather like a Unicorn, one thinks), and a long hairy tuft of red feathers on the back of the head. These latter would make excellent salmon-flies, and indeed the bird is especially hunted for that very purpose. The rest of its plumage in the male is a rich green, the wings brown, mixed with buff, and the under-parts black.

Here, a hundred or so feet from the top, my native collectors found the young of a very rare bird indeed, *Eupetes macrocercus*. Mr A.H.Everett first found this bird on Mount Penrisen, and it has also been found by me on Mount Dulit at a height of 4000 feet. It seems clear that the species inhabits only the highest points, and lives an isolated life in those places. The young birds differ considerably from the adults, being much duller and browner in colour, while the crown is reddish-brown instead of bright chestnut. The throat is white and the under surface a slaty black. The sexes appear to be alike in colour.

Just before nightfall we moved our camp to the

actual summit, a curious dome-shaped top entirely covered with moss. Beneath our feet it lay in a thick carpet, so that to find our way through we had to crawl on our hands and knees; above our heads it formed archways, while every tree was inches leep in it. There was moisture, too, everywhere; everything we touched was wet, and it was clear that we could not stay there any considerable time in any comfort. On the other hand we had reached our journey's end, richer in experience and in specimens; and were now looking down, recognising familiar landmarks, across some of the thickest tropical jungle in the world.

Note.—For the description of the dense fibrous tissue of the scalp (page 155) of *Pityriasis gymnocephala,* I am indebted to R. H. Burne, Esq., F.R.S.

CHAPTER VII

THE WEALTH OF THE JUNGLE

Mangroves and Cutch—The Nipah, a "Universal Provider"—A Prehistoric Woolworth's—Sago—*Lamanta*—The Punans—Camphor-hunting—Tests for Crystals—The camphor-language—Gutta-percha—Wild Rubber—Gums—Copal and Dammar—The Rattan—Black and White Pepper—Rice—The Kalabits—A Willing Victim of Alcohol—A Kalabit Salt-factory—Chinese Influence in the Hinterland—An Ingenious Scarecrow.

AMONG the many noteworthy features of the great island of Borneo (and of Sarawak in particular) not the least are the natural products. The most prolific oil-field in the whole of the British Empire is that of Miri, a few miles south of Baram Point and the Baram River; but, to come to smaller matters, the true voice of Romance itself is heard in the names Cutch, Antimony, Sago, Gutta-percha, and Camphor, many of them articles of everyday use, but whose provenance is mysterious and arresting.

One has not far to go for one of the chief local industries, for as the steamer that plies between

Singapore and Kuching, the capital, passes slowly up the Santubong branch of the Sarawak river, the eye is caught by the dense masses of dark green tree-growth right down to the water's edge, the roots appearing to rise from the very stream itself. These are the well-known Mangrove trees, whose bark gives us the valuable tanning agent known commercially as Cutch; on either side of the river the forest is lined with them to a height of some ten or twenty feet, a continuous hedge of the most forbidding aspect. The trees themselves, the foliage of which is in colour much like that of a laurel, look as if they were suffering from some form of locomotor ataxia, and had lost all control of their members; for the roots, starting from the trunk itself, are above the surface and spread out in every direction like the tentacles of an octopus. From these main limbs smaller roots branch out, until the whole forms a ganglion, as Dr Beccari says, " monotonous, weird, and desolate." The fruits of this strange growth are pantomimic sausages about a foot long; but what is more extraordinary is that from the centre of the leaves, on the smaller branches there stretch out " fusiform " appendages varying in length from two to eighteen inches, and sometimes as thick as a walking-stick. These are the roots of daughter trees, for the seeds in the fruits germinate while they are still attached to the parent plant, and before falling to the ground. The resulting exuberance of confused vegetation may be imagined; while under the branches all is in the deepest and most melancholy shadow, wherein mosquitoes make life intolerable for more than a few minutes at a time. The only relief to the gloom is given by the many beautiful orchids hanging from the branches, some of them of the most delicate shades, and other plants not unlike the flower of the Rhododendron.

There are three varieties of Mangrove found in Sarawak, their local names being *Putut, Bakau,* and *Tengah.* The first two of these are the true mangroves, and are both used for the extraction of cutch, Bakau being the better of the two. Neither of them, however, is anything like so valuable as the Tengah, which is perhaps not a mangrove at all, and is very much rarer. It seems, however, to have been known to the Chinese for centuries, long before any European set foot in Borneo; for at the time when the Chinese worked the gold at Santubong (which may be any time as far back as two thousand years ago) they appear to have used the tan for their fishing nets; while two or three centuries ago people from Pegu and Upper Burma were working there, as we learn from certain old coins. Europeans do not seem to have done anything with it until the industry was started in 1897 by Mr H. H. Everett; in later years it has been taken up seriously by the Island Trading Company, several tens of thousands of pounds' worth being exported annually.

It will be easily understood that work in these dismal swamps is likely to be highly unpleasant and of doubtful profit, but according to the statement of those best entitled to speak on the subject, " Only those who have practical knowledge of the characteristics of the vast mangrove swamps of the Tropics can form any conception of the difficulties attending the collection of mangrove bark." The work of collecting and treating is all done by natives, except for some few European managers and engineers.

The particular value of the Cutch tan is that it

can be used to produce a lighter shade than any other agent, the finished colour of the leather often being a pale buff. For this reason it commands a correspondingly high price in the market; a box of one hundredweight being worth anything from twenty to twenty-five shillings.

To obtain the extract the bark is first stripped off the branches and boiled, and then, after being evaporated, crushed under heavy pressure, when it gives out an extract which looks and smells very much like strawberry jam. This extract, after being cleared of sediment, is put into centrifugal machines and whirled round until it attains the consistency of toffee. It is then packed into boxes or bags; after which it hardens on exposure to the air until it finally looks like dark red sealing-wax. The boxed extract is used for the tanning of fishing nets and sails, and for dyeing; its value lies partly in its variation of shades, and also on its durability and the fact that it contains no sticky sediment. The wood of the mangrove is used as household fuel, and most of the firewood in Kuching comes from this source.

In those parts of the coast where one finds mangrove swamps are also to be seen three of the most useful of Bornean palms, the Nibong, Nipah, and the Sago. The Nibong is a tall tree with a straight slender stem, crowned with a tuft of delicate drooping fronds. It is used extensively by natives in house-building; the trunk makes enduring piles and beams for the framework, and, split into slats (*lantai*), serves for the flooring of both houses and boats; it is also used for the darts used in the blow-pipe. Its usefulness is, however, far surpassed by the Nipah, which for the Bornean is a sort of "Universal Provider." This, like the mangrove, grows between land and water, in swamps; in appearance it is rather like a stemless coconut palm. As a matter of fact it has a very thick heavy stem, but this is so much hidden away behind the foliage as to be almost invisible; it is, besides, squat and flattened and throws off a number of roots or suckers from the base. The head of the palm is, therefore, never at any great height from the soil; but the fronds, which are delicately fringed so as to resemble the "hatchings" used to represent mountain-ranges on maps, often reach the height of thirty feet. As the Nipah is most prolific in growth, it forms a beautiful but somewhat monotonous barrier between the stream and the jungly hinterland.

The Nipah palm has rather striking yellow flowers, the stems of which give out a juice from which sugar is made by boiling; when fresh, this juice is sweet and refreshing, but after being kept it ferments and turns to vinegar. The amount of juice thus obtained is little short of miraculous; from one flower-stem alone in a single night as much as three quarts can be obtained. This is collected by means of a sort of funnel of jointed bamboo.

The fruit of this useful tree, in a way, resembles that of the coconut, for it possesses an outer cover (from which fibre is obtained) and, inside, an opaque " milk " fluid and also an albuminous lining to the shell. Here, however, the resemblance ceases, for the Nipah " nuts " grow in clusters of forty or fifty, each fruit being about two and a half inches long, and the whole bunch about a foot across, and somewhat resembling an enormous flower. The " meat " hardens on exposure to air, and provides a very useful sort of vegetable ivory, from which small articles such as buttons and studs are made; care, however, has to be taken in the obtaining of the material, for from the "meat" springs the germ of the young plants, which contains a large amount of oil; and unless the "meat" is properly dried it is useless for commercial purposes.

The stem, which can only be called wood by courtesy, consists of a great porous mass encased in a skin; it is one of the few parts of the tree which are not commercially valuable, yet even from it, when it has grown old and has been burnt in a fire, a certain amount of salt is obtained.

As for the leaves, they are put to dozens of uses. Bags are woven out of the young green fronds; the grown leaves make "kajangs"; the young green leaves, when stripped of the outer skin, are used in the place of cigarette-papers or covers for the local "rokos" or cheroots, and the fronds for the manufacture of baskets, wrappers for packages, roofs for houses; from the leaf-ribs are made brooms, stands for cooking-pots, traps, and small household articles.

Evidence from fossils shows that during the

Tertiary geological period this palm was found in the south-east of England; Doctor A. B. Rendle, in the Journal of the Linnean Society (*Botany,* Vol. xxx, 1894), describes a fossilized fruit found in Sussex; this is a fact of some importance, since the Nipah is to-day an inhabitant of the tropics exclusively. It is interesting to think that this "Universal Provider" anticipated Mr William Whiteley, or should one say Woolworth?

The Nipah palm is represented at the present day by a single species, *N. fruticans,* restricted to the muddy estuaries of the Indo-Malayan region. A dense crown of tall "feather" leaves fifteen feet or more in length springs from a thick prostrate root-stock; the fruits, each containing onc seed, are crowded into a large round head borne on a woody stalk. (The lower figure represents a head from Singapore, and above is a single fruit.) In later geological periods the palms reached much farther north than at present. In the London clay at the mouth of the Thames and in similar and closely allied beds at several places on the south coast, as also in France, Belgium and Italy, are found the *Nipadites* fruits, which show a much wider range in specific forms than the modern representative. (See Rendle in Journal of the Linnean Society (*Botany*, Vol. xxx, p. 143)). The figure shows a fruit of *Nipadites Burtini* from the beach at West Wittering, Sussex, where a layer was exposed in November 1890. These fruits were empty shells, suggesting that they had germinated *in situ* and the shell had subsequently become filled with the sand. No trace of stem or leaves has been found.

The third of these swamp-growing palms, the Sago (*Metroxylon*), is one of the most characteristic products of the island, for Sarawak is one of the principal sources of the world's supply of sago; and, according to some scholars, the Malay name for Borneo, Pulu-Kalamantan, means Sago Island.

Sago is, of course, a starch prepared from the pith of the palm, which itself is a large tree with a thick trunk, and from twenty-five to forty feet in height. Like the proverbial aloe (or *Agave*), it blossoms only once, and then dies when the fruit is ripe; great care is therefore taken to prevent the tree from putting forth its flower-spikes, for otherwise there would be no sago, the starch being all used up in the action of flowering and fruiting.

NIPAH PALM FRUIT AND FOSSIL OF THE SAME

There are at least two kinds of cultivated sago palm in Sarawak, one of them being distinguished by a strong growth of thorns round the base, which keeps off wild pigs and other depredators, but is most unpleasant to the bare feet of the workers. In both kinds the sago grains are fully formed in a healthy tree at about the age of six to eight years. The best results are obtained from seed of young plants; when a tree is cut down new shoots grow up from the stump, first running along the ground like strawberry-runners, and then rising vertically and independently without cultivation.

Some of the best cultivators of sago are the Milanos of the coast, the part where my friend Boom lived; they produce quite a fair percentage of the total supply of Sarawak, for they prefer sago to rice as their chief diet. To obtain the marketable product they fell the tree, and cut it into logs about three feet long, which are then split lengthwise, and the pith is either pounded out with a mallet, or else scraped out with a sort of rasping tool. This is made out of a board about five feet long into which are driven, at intervals of half-an-inch, long wire nails in such a way that the points protrude an inch or so on the other side. This is worked to and fro by two men, like a cross-cut saw, by means of handles fastened at either end. It is extraordinary how quickly it frays out the pith, which is thus easily collected in mats and placed on a bamboo platform with a large trough below it. Running water is led from a bamboo pipe to the platform and the sago grains are stamped out by the feet and pass through the mats into the trough below.

When the washing process is partly done, the raw sago-grains are made up into leaf-packages called *tampins,* and are either shipped at once to the factories (of which there are several, at Kuching and elsewhere), or else are kept in a damp condition. The *tampins* are made from the leaves of the sago tree, and are tough and durable; but a certain amount of fermentation takes place, making the whole atmosphere reek with the smell of sulphuretted hydrogen and decaying vegetable matter. (It is at this stage that the word *lamanta* is used for the sago). At the factory it is filtered through coarse matting into tubs—the process being repeated three times in as many days. Finally

(Upper) CULTIVATED SAGO-PALMS IN A SWAMP
(Lower) PUNANS WORKING FOREST SAGO.

it is dried and shipped in bags in the form of a beautiful white flour, to Singapore, which is the chief exporting depot.

Wild sago is not much eaten by the majority of the pagan tribes, such as the Kayans and the Kenyahs, who use it only when their rice-crop has failed, or when from some other reason there is a shortage of *padi*. The Punans, however, eat no rice at all, but substitute for it this wild sago (known to them as *nanga*), which is easily worked by splitting a section of the bole of the sago-palm and then pounding the pith out with a wooden mallet on to mats. The mats are then put under a small fall of water and the starchy grains washed out of the pith through the interstices of the mats until it settles down at the bottom of the trough in a firm white mass. After several such washings the grains look very much like fine silver sand.

The jungle-sense and practical forest-lore of the Punan are, in any case, exceptional, but on one subject he may be said to be an expert. This is Camphor, a product not difficult to obtain, but requiring special knowledge and special care.

There are on the market to-day two entirely

different kinds of Camphor. The ordinary variety, which is found in China and Japan, but especially in the island of Formosa, is known to the chemists by the name of Borneol—I imagine from the mistaken notion that it comes from Borneo—and is obtained from a small tree of the laurel family, Cinnamomum Camphora. Bornean camphor, on the other hand, is indigenous only to Borneo and Sumatra, and, possibly, some of the smaller islands; it is found not in a shrub but in a very tall tree, which is also quite useful and valuable as timber, and is very much more valuable than the Formosan kind. This may be seen from the fact that in China the price of ordinary camphor is about a shilling a pound, whereas for the Bornean, if the crystals are of a good size, as much as £4 per katty (about 1⅓ lbs.). The Chinese are great users of camphor, for various purposes, but specially for embalming; the Great Empress (" The Old Dragon "), who died in 1908, required over £30,000 worth of camphor for this purpose. It will be easily understood, therefore, that a good synthetic camphor, equal to the best Bornean, would find a ready market and a profitable one. I myself at one time attempted to produce some, using Borneol as my basis; however, when I took it to the Chinese experts in Borneo, they smiled and said, " Oh, yes! This is very like the Bornean Kapor! But does it respond to the lemon-juice test?" (Of course they had a test different from anyone else's, but a remarkably good one). " Where does this come from? It may be worth a lot in its own country, but it is not the same as Borneo Camphor!" They understood that it was neither Bornean nor Formosan, and what they wanted was the Bornean, and no other.

The Bornean Camphor-tree (*Dryobalanops aromatica*) is, when fully grown, at least two hundred feet in height; down to about five feet from the ground it is straight-stemmed, but from that point it gradually broadens towards the roots. It is between this point on another, about seven or eight feet higher up the stem, that the camphor crystals are found. Camphor hunting, however,

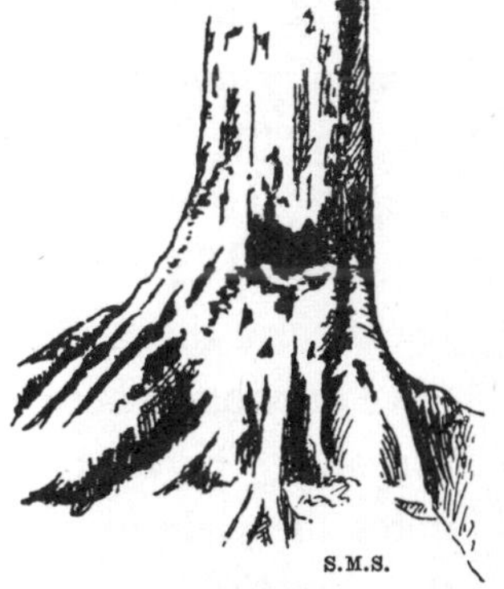

CAMPHOR TREE SHOWING HOLE

depends for its success upon a knowledge of the individual tree and its condition; for unless the tree is exactly "ready" for felling, the labour is merely lost.

To discover whether this is the case or not, the Punan's procedure is both methodical and truly scientific. He first selects a tree which, by its size and apparent age, is likely to prove productive. Next, he observes the leaves, which, at a certain stage, exude a sort of oily or soapy gum; and, as a further test, cuts a deep notch in the bole of the tree with a long thin axe. He then widens and deepens this hole until it is about a foot deep in the tree, and a foot in height. The outer shell of the tree, which is usually hollow, is now half cut through, and by smelling the chips of the wood the Punan can judge, roughly, at what stage of formation the crystals are. If oil of camphor comes out, he puts a leaf-bucket or a tin can into the hole (the walls of which have previously been burnt with fire), so as to catch the oil. The next day he returns, and carefully notes the yield of oil; if it is abundant, he decides that the time has not yet arrived for felling; but he notes the tree for future use, because he knows that in a year or so the crystals will have formed.

A later stage in the growth is inferred from the presence of gum and a putty-like substance, rather like a soapy dough, where there was formerly oil. This exudation is known to the Punans as *Paji buta* or *Kapor buta* (blind camphor).

The third stage is when there is very little oil except on the outside of the tree. It is now ripe for felling, if it passes the other tests, *i.e.* the smell of the wood and the age of the tree as judged from the hollowness of the sound emitted when it is struck. The tree has, naturally, been under observation for some considerable time and has been noted as such —for the Punan, in the utterly trackless jungle, has little more difficulty in locating his trees than a London taxi-driver has in taking you to an address in Bayswater. Moreover the cutting of the trunk seems, so far from harming the tree, to do it good; for the camphor-crystals never occupy the whole of the trunk, and there is always a sort of flue in the middle of the tree. This flue seems somehow to aerate the tree, and I am inclined to think that it is only when such aeration has taken place that crystals can be formed. Camphor contains oxygen in its composition, the chemical formula for Borneo camphor being $C_{10}H_{18}O$., at least in its crystalline form. The mystery which the Punan has solved for practical purposes still remains for Science to elucidate; but it seems to be the fact that the Camphor tree, like certain plants, is capable of absorbing oxygen from the moisture of the forest, and from the air.

When Punans go to look for camphor, they are both methodical and ceremonious, and many restrictions have to be observed. In the first place they will not work for more than a certain number of days in a year (incidentally, it may be observed that they reckon by nights spent away from home, rather than by days of work); next, they may speak to no one; if anyone calls to them they refuse to answer. Further, omens have to be taken with due care, by the observation of certain birds from booths specially built for the purpose. Here they sit and watch for the appearance of auspicious birds, meanwhile smoking native cigarettes, the smoke of which is supposed to carry their wishes and prayers up to the gods. All this time, no word must be uttered to any outside party; the whole procedure is, perhaps, partly based on commercial keenness; but there is underlying it a feeling of a Life existent in what to us seems inanimate, or anyhow soulless, creatures, such as trees and creepers. They must not speak about the camphor-trees, for instance, for fear lest the spirit in the trees should hear them, or be informed by the birds, and their expedition prove a failure. On the same principle, if their journey is up the branch of a river, they put up signs on a strip of rattan hung across the stream, so as to make it *mali,* i.e. to indicate that it is consecrated or set apart for a religious or semi-religious purpose. The act is, of course, a form of Taboo, the Punan word used being *pantang*; but it also indicates the work on which the hunters are engaged.

If a stranger, or a member of another community, accidentally or otherwise, breaks through this restriction, and speaks to the Punans, he realises when they do not answer him that they are engaged on some ceremonious business; but if a word slips out from a Punan, the party breaks up; for they say " the camphor will evaporate "—as a matter of fact camphor does evaporate on exposure to air. Another curious restriction is that if camphor has been obtained it must not be shown. It is kept wrapped up in palm-leaves or bark-cloth, which in its turn is enclosed in an airtight bamboo case. What, however, is perhaps the most extraordinary feature of the whole procedure, is that a special language is used throughout; not only may the Punans not speak to any outsider, but when they speak to one another they use a tongue which is neither Punan, nor Kayan, nor Malay, nor any other known language, but may perhaps be an older tongue of which the modern Punan language is a descendant. They want some means of communication which no one but themselves can understand; their use of it is like that of Thieves' Latin, or Back-Slang. Some of the camphor-hunters of Johore (the Orang Benua) have a similar language, and also some of the Milanos, whose language corresponds to some extent with that of Johore. From this it would seem as if camphor was originally collected by Malays; certainly in all the three languages there are Malay traces, and it is possible that the Punans may have taken the idea

KAPIT FORT; A SARAWAK OUT-STATION FROM THE AIR.

of using a non-Bornean tongue from their dealings with Milanos and Malays—very much as in English slang a number of Romany words (*e.g* " tanner " and " pal ") are used by people who themselves are not Romany Chals or Romany speakers.

A typical forest product of Borneo is Gutta-percha, the name of which is Malay, Percha being Malay for the island of Sumatra. This, like caoutchouc, which it much resembles, is the *latex* or milky juice of a tree fairly common both in the coastal plains and alluvial deposits, and in the higher forest of the interior. The tree is a tall slender one, its diameter when fully grown being about two or three feet. The latex is obtained by felling the tree and then ringing the trunk at intervals of about eighteen inches, a leaf-cup being placed under each ring for the reception of the juice. This juice, when fresh, is white in colour, but it darkens on exposure to the air and takes on a tinge varying from pale yellow to dark brown. At an ordinary temperature Gutta has certain of the qualities of leather, but is not so flexible. When heated to the temperature of boiling water it becomes soft and malleable. Its chief use is as an insulating and protecting material, for it is a very poor conductor of electricity; it is used largely for cables, and, as an older generation will remember, until the arrival of the "Haskell" in the early years of the present century, supplied the world with golf-balls.

Three other varieties of wild rubber are found in Sarawak, the commonest being known as *Jelutong,* an inferior product but very plentiful in the low-lying forests, and very easy to work. The tree is one of the largest in the jungle and can be successfully tapped without any damage being caused to the tree. Other species, not so common as Jelutong, are known as *Kati* and *Plai.* The latex is treated simply, being made up into large round blocks, which are kept floating in water until sold. Of late years a considerable industry in this inferior rubber has developed, as it is largely used in the composition of American chewing-gum.

Besides these, there are at least two other kinds of wild rubber, known to the Ibans as *Kubal,* and to the Kayans and other tribes as *Pulut.* These wild rubbers are creepers, which attain a considerable girth, sometimes as much as six inches or a foot in diameter. The latex is collected in the same way as gutta, the creeper being cut down, ringed at intervals of some eighteen inches or so, and the juice collected in leaf-cups and then transferred to larger receptacles made of bark. The variety known as *Kubal susu* is far superior both in quality and output, the other kind, known as *Kubal Arang* (Arang meaning coals), giving a juice which goes black on exposure to air and is of such poor quality as to be hardly worth working. The viscid mass, when collected, is poured into a large open iron pan (*quali*) and is boiled in water to which a handful or so of coarse salt is added. When this is done the latex at once congeals in large lumps. Before it has time to cool a hole is pierced in each lump and a rattan passed through for convenience of handling. It is then cooled off and stored in cold water until sent to the market. This rubber, which is a true Rubber (*Willughbeia flavescens*), is of good quality, quite white and spongy in appearance, and when bounced on the ground is very resilient. The fruit of the creeper is luscious and is much sought after, and the seeds, which are swallowed whole, are about the size of a plum-stone, but not hard. The creeper itself attains its maximum girth between the age of five and ten years, according to the suitability of the soil in which the seed has fallen.

Many of the Bornean forest trees contain large quantities of gum, which usually exudes from wounds or injuries to the trunk and bark. These gums vary very much in appearance and quality, some being reddish and opaque, others yellowish-brown; but the finest qualities are almost as transparent as glass and are known as *Mata Kuching,* or Cat's Eye; these varieties are comparatively rare, being produced by one or two kinds of tree only. The commoner sorts, known as *Daging* (meat), are often found in quite large masses, while the commonest kind of all is that which, curiously, is picked up along the sea-coast, having been thrown up by the waves from trunks which have been carried down the larger rivers and so out to the sea. Before Kerosine oil was introduced into the country by traders, the peoples of the interior had no other form of illumination for their Long House, apart from such light as came from the communal log fire; but in every room and in the gallery of the long house could be seen gum torches, and can be seen to-day in the remoter villages. These gums are also used for the caulking of the seams of boats, and for other less important purposes, the commonest variety being known as *Impernit,* the product of a tall forest tree common all over the country, and growing to a height of over two hundred and fifty feet.

The finest qualities of Bornean gums are those known as Gum Benjamin and Gum Copal, which are largely used in the preparation of varnish. In Sarawak of recent years a considerable trade has sprung up in these gums, under the name of Dammar.

The most characteristic plant of the whole of the Malayan region is, however, the Rattan (rotan), of which there are, in Borneo alone, at least a hundred varieties. These palms, the botanical name of which is *Calamus,* are unlike many other members of their family, in that they are creepers, which, in their efforts to reach the sunlight, climb up the stems of the tallest forest trees, catching hold of the branches and foliage by means of their thorns and spikes. I have myself seen one of these palms cut down which measured over two hundred feet in length. For commercial purposes they are cut into lengths of about fifteen feet—*i.e.* two and a half spans of a man's arms from finger-tip to finger-tip—and after being dried are floated down the rivers on rafts ready for sale. The varieties in greatest demand are the two kinds of *Rotan Saga* (one large and the other small) and the *Rotan Gamaing* and similar kinds; the well-known Malacca canes are also of the Rattan family. Their uses are too well known to need description; but the canes are not the only useful product, for from the ripe seeds of the *Rotan Saga* a deep red dye is obtained, known commercially as Dragon's Blood.

A Bornean product of some importance and of profit to the cultivators, is pepper, the growing of which is almost entirely in the hands of Chinese. Pepper is, of course, the berry of a vine of the same family as that which produces the betel leaf used in conjunction with areca-nut for chewing. The pepper plant, which is fairly common throughout the East Indies, seems to have been introduced into Borneo by Indians, but its cultivation was very early taken up and extended by the Chinese. In

1876 the Rajah of Sarawak, in conjunction with certain Chinese merchants, laid the foundations of the industry, until Sarawak pepper gained for itself the reputation of being superior to any grown elsewhere. It had its ups and downs; there have been periodical booms and falls, but almost throughout Upper Sarawak proper, *i.e.* the land drained by the upper waters of the Sarawak River, prosperous pepper gardens are to be found. The rainfall in this part of the country is about 130 inches per year, two-thirds of which fall during the North-East Monsoon; the soil is for the most part a stiff yellow clay, which exposure renders friable; this forms an excellent base for the main roots of the vine, and also, when burnt, makes a very useful top-dressing. The Chinese are experts at making burnt earth, a task which demands both skill and patience.

The pepper vine is invariably propagated by means of cuttings. These cuttings are usually from eighteen inches to two feet in length and are put in the soil fairly deep, and at an angle of about 45°, leaning towards light sticks on which they are to be trained. From the first two or three months, and until the cutting has begun to shoot, they are protected by fern-leaves and grass cuttings. At periodical intervals (in some cases of a month, in others of three or four months) they are banked up with burnt earth as one does in growing celery. Later (usually after eighteen months) ordinary or "raw" earth is substituted. As soon as is necessary the young shoots, which are planted in groups of three, are tied to the sticks with soft bark or twine; then, after about six months, permanent posts are set up. These are made of the hardest wood obtainable and are usually about twelve feet in length. The vines are tied at intervals of about four inches, more or less at each joint in the stem.

Fruiting, in the ordinary course of events, takes a year; that is to say, within a year the crop is finished. The first crop is obtained when the vine is about two years and a half old. The average life of a vine is from ten to twelve years.

The main cropping time falls between July and October; to obtain the marketable product, two methods are used, according as white or black pepper is required. To make white pepper, the spikes of fruit are picked just as they are beginning to turn red. They are first crushed under foot so as to loosen the berries from the stalks. The whole is then put into bags and soaked in water, slow running water being prepared because it gets the full heat of the sun and therefore assists the decomposition of the skin. After a week or so, when the skins are sufficiently loose, the pepper is put into tubs and stamped out until all the skins and stalks have been cleared. The peppercorns are then spread on mats and dried in the sun.

For black pepper, the fruit is picked not quite as ripe as for white pepper, and is generally simply dried in the sun, the skin drying on to the peppercorn and turning black. Later the pepper is rubbed by hand so as to separate the berries from the stalks, which are then winnowed out.

The whole process is very simple and not very expensive; very little labour is needed, for one coolie can attend to as much as four hundred vines. Disease has to be guarded against, but the Chinese have their own specifics, such as decaying tobacco and a solution of Tuba. The profits, on the other hand, although slow, are fairly sure and not inconsiderable. The whole industry is thus one for which the Chinaman is admirably suited, and which admirably suits him.

Pepper is, however, not the only agricultural industry in which the Chinese have taken an important part; they have also been, at least in certain parts of the island, the pioneers of rice-cultivation, the wild rice (*Oryza sativa*) not being indigenous to Borneo.

The successful raising of rice-crops means life or death to the majority of the people of Borneo, for there are only a few tribes (such as the Punans and other " food gatherers " and the sago-growing people) who do not make the cultivation of rice their most important work.

Whether the Chinese did or did not introduce rice into Borneo, is an unsolved problem, and one of no great moment. What is however significant is the improvements that they have effected. The Dusuns of British North Borneo, who are perhaps the only natives to use a plough, learnt that art from the Chinese; and, in fact, many Dusuns have Chinese blood in their veins. What is even more extraordinary is to find Chinese influence in one of the remotest part of Central Borneo, in the country of the Kalabits.

The Kalabit country (which also contains a number of Muruts) is a high tableland from three to four thousand feet above sea-level and situated on the watershed between Sarawak and Dutch Borneo—*i.e.* in the north-east centre of the island. The Kalabits are a branch of the Murut group of tribes, and very probably came to Borneo from the Philippines or from Annam. They are by no means a beautiful set of people, the women in particular being of a degraded, sensual and even brutal type. Whether for this reason or not, they have a peculiar custom, in that it is they who propose marriage to the men, and not vice versa. Perhaps if the women were more attractive, the need would not arise.

The Kalabits, apart from anything else, are very capable cultivators, and it was from the Chinese that they learnt the art and practice of agriculture. They and other Murut peoples, alone of the Pagan Tribes, cultivate rice in terraces and with a system of irrigation, and they alone raise two rice-crops every year. It is not at all improbable that the Chinese introduced various kinds of padi, and they may have taught the local youth the uses of rice-spirit. It is a fact that Kalabits and Muruts drink more rice-spirit than any of the other tribes, and it is not uncommon to see a half-intoxicated man recumbent on his mat by the side of a capacious jar containing his favourite beverage, and taking occasional sips (or perhaps swigs) through a bamboo tube running through the jar from bottom to top among the fermenting rice. The whole apparatus has been devised with a skill that argues long experience and an ingenuity worthy of a better cause. Inside the jar a grating made of bamboo slips is placed, on which the fermenting rice rests and through which the spirit slowly trickles. The bamboo tube works like a syphon, and the willing victim of "Old Man Booze" or "The Demon Rum", as the Americans call it, literally takes it lying down.

The Kalabits are, however, not a degraded people, as may be judged from their efficiency as farmers; there is much of interest in them and in their country, and it was much to my regret that I was unable to explore their part of Sarawak. One expedition, indeed, very nearly cost me my life, for both my own boat and also that of the principal chief who accompanied me, jammed together in some more than usually awkward rapids, and were both smashed into smithereens. I myself was hurled out of the boat, washed down a respectable-sized waterfall and bumped severely against a large rock; when I was at last helped out by one of our party who had been more fortunate and had succeeded in reaching a place of safety, I found I had been carried a good fifty yards down a very violent reach of water. As we had lost a considerable amount of stores, and the two boats were shattered beyond repair, there was no help for it but to turn back; and I never again had the opportunity of making an expedition to these people. Some years after, however, Mr R. S. Douglas, who succeeded me as Resident of the Baram Division, led a peace-making expedition into the heart of the Kalabit country; and an interesting account may be found in the journal of the Sarawak Museum, Vol. I, No. 2.

Mr Douglas on this occasion investigated the local salt-springs and salt factory. I had come across this salt myself, for it had a widespread reputation and was exchanged with the neighbouring tribes for weapons and rubber. It had a reputation as a preventative of goitre, and I suspected that this belief was well founded, for Kalabits were not subject to goitre, whereas their neighbours (Kayans and Kenyahs) on the other side of the limestone range which borders the Kalabit country were commonly sufferers from it. I had some analyzed in London, and it showed the presence of iodine and lithium.

Mr Douglas has described the process of making the salt. The water from the springs is first collected in large flat cauldrons, which are heated over enormous fires; during the boiling process the salt coagulates round the brims and is scooped into bamboo vessels. These, when filled, are sealed against the air, and after a time burnt in furnaces, from which, the bamboo having been burnt off, the salt appears hard and white and of a long cylindrical shape. Kalabit children are given pieces of this salt to keep them quiet, as other children have lumps of sugar.

As for the Chinese in the Kalabit country, it is of great interest to note that the present District Officer of the Baram, Captain Andreini, has recently come across archæological remains of some importance, in particular a stone monument, about twenty feet high, with an inscription in Chinese in which the words "Sum Peng" have been deciphered. Ong Sum Peng was the Chinese governor of a colony on the river known as Kina Batangan (the name of which, in itself, shows the Chinese connection); his sister married the first Moslem Sultan of Brunei. During his period of office (which was probably nearly a thousand years ago) Chinese merchants had quite a noteworthy amount of trade with Borneo, exporting pepper, camphor, edible birds' nests, tortoise-shell, dragon's blood and other commodities.

The Chinese have left a number of unrelated indications of their presence in Central Borneo, among others in the name Kina Balu, the highest mountain in the whole country; this name, which is usually supposed to mean "the Chinese Widow," much more likely is a Chinese form of the Malay "Kina Bahru," *i.e.* New China.

Rice is, then, the most important food-stuff of the native peoples; its cultivation varies with varying tribes. The Dusuns of the north, as I have said, alone use the plough; in other parts the fields are roughly churned up by buffaloes. The Kayans are the most systematic, the Kalabits and Ibans the most industrious, and the most scientific. A labour-saving device which they use for scaring birds might well be adopted in countries which are commonly supposed to be culturally more advanced. At various points in the rice-fields are set up posts supporting lines of twine or rattan, on which are placed bamboo rattles, tins, sea-shells, bits of metal, and various other contrivances which give out a sound when the string is pulled. All these cords are linked up together and connected with a larger rattan, the movement of which starts the whole belfry. This "control-handle" is extended and broadened out like a fan, and is then attached to a stout pole set up in the mid-stream of a river. The stream acts on the fan as on a turbine, and causes an oscillation of the rattan "joy-stick," which carries on the good work to the strings between the posts. Thus, if there is any sort of head of water, a constant jingling is set up which effectively keeps away birds and other marauders.

CHAPTER VIII

SOME MINERALS AND POISONS

Antimony (Stibnite)—The Earth Spirit Outraged—Retribution—Poisons—The Upas Tree—An Antidote to Dart-poison—Derris Elliptica—Datura—The Papaw—The Champak—The Caladium and Cuckoo-pint—Fish and Insect Poisons—A Snake Story—The Uses of the Medicine Man.

MUCH of the mineral wealth of Borneo has never been investigated; and, in view of the dense primeval forest which covers so much of the surface, it will doubtless be many years before its potentialities are discovered. The Miri oil-fields, which have been mentioned above, is an exception; and there is undoubtedly a great deal of coal, which has never yet been profitably worked to any great extent. The Chinese at one time made a profit out of gold; and in the rivers there is undoubtedly a certain amount of alluvial gold. Evidence of the existence of diamonds has been found, and at a place called Tegora there is a valuable deposit of quicksilver in its "native state"; in the form of cinnabar it has yielded a certain amount of profit, and it is quite likely that at some time antimony will again be worked commercially. This metal is usually found in the form of Stibnite, or Sulphide of Antimony, and has been located in several places in the Baram District in the form of an outcrop. It is found either in large nuggets (which are the most productive) or else in beautiful long spear-like crystals from nine to twelve inches in length, or as Native Antimony.

In colour Stibnite resembles pewter or unpolished silver, from which fact it gets its native name *Batu Perak* (silver rock). Round the Tisam river (a tributary of the Tinjar) a considerable surface outcrop was discovered some years ago by Mr H. H. Everett, who worked it for several months during the fine season with a gang of Malay miners, for the Borneo Minerals Company. This was merely a surface outcrop, or what is called a "pocket"; but it is quite possible that deeper excavations would give a considerable yield.

The work connected with the exploitation of this deposit was not without its incidents. In the first place there was a strong feeling among the local tribes against interference with the soil. The old

(Upper Left) ULAT BULU CATERPILLAR
(By permission, Borneo Company)
(Upper Right) STICK-LAC INSECT.
(Lower) STIBNITE CRYSTALS.

Sebop chief, Aban Jau (who was well known in Sarawak at one time as a relentless enemy of the Government, but who later became one of the staunchest adherents of the late Rajah, and whom I counted it a pride to be able to number among my personal friends),* had, years before the work started, done his best to prevent any interference, as he considered it, with the Earth. He predicted that such operations would disturb the Earth Spirit (*Hantu Tana*), a deity which among the Sebops is considered only next to Laki Tenangang, the Supreme Ruler of the Universe. With the passion of conviction, he assured us that the outrage which we proposed would certainly be attended with disaster.

During Aban Jau's lifetime no actual operations were carried out; but some time after his death, Mr Everett, acting on my advice, started working the outcrop. To the horror and amazement of those who remembered the old chief's warning, the enterprise was accompanied by disaster. Several of the Malays and others engaged in the work died of Beri-beri, including a half-caste foreman named

* Readers interested in the history of this fine old chief are referred to my "Fifty Years of Romance and Research," Chap. III ("A Rob Roy of Borneo").

Lopes. I could hardly help thinking that the disturbance of the soil might be, in some obscure way, connected with our calamities, more especially because, at that time, practically nothing was known about the causes of Beri-beri.

The second incident was hardly so tragic, and was more capable of a rational explanation. The amount of Stibnite collected (a few tons) had been piled up in a heap near the bank of the river, so that when required it might easily be transported into boats; but, between the collection and the removal, various delays occurred, so that a year or more elapsed before it was removed. When Mr Everett and myself went up river to arrange for its transport, we found that quite a noticeable amount had disappeared entirely, while round about the dump were heavy pieces of the ore, detached from the main body.

Explanations were, of course, not lacking; and there were plenty who said " I told you so ! " Some suggested that the detached pieces were trying to get back again to their original home; others hinted that it was the work of the spirits of the ancestors of the villages, and it was remembered that none of the local people had ever been allowed to take any wages for what work they did, and that they had been discouraged in every way. A little local investigation showed the real cause—wild pigs; for their tracks and those of other animals were visible in many places. One knows that tame pigs in England very often eat cinders, and it is probable that the stibnite had been taken as some form of medicine, perhaps as a remedy for swine-fever.

Among the many vegetable products of Borneo (though they cannot accurately be termed economical products) are poisons of various kinds. Of these the best known is the tree called in Malay *Ipoh* (which however is, strictly speaking, a generic term meaning any kind of poison), the Upas tree of fable, the " Hydra-Tree of Death," as Erasmus Darwin called it in his " Loves of the Plants." This tree, whose supposed qualities have become proverbial, was for years alleged to give off poisonous fumes fatal to animal life. The legend seems to have started from an account given by a Dutch doctor in the eighteenth century, who wrote that criminals condemned to death were offered the chance of life if they would go to the Upas Tree and collect the poison. Nine out of ten, he said, perished, for "not a tree nor blade of grass is to be found in the valley or surrouding mountains. Not a beast or bird, reptile or living thing, lives in the vicinity." It is hardly necessary to say that the yarn has no foundation whatever, and the description would almost seem to stultify itself; and subsequent travellers have proved clearly that it is no more unhealthy than any other forest tree; and as for the poison, the juice of the tree can be applied to the unbroken skin or taken internally without harm.

On the other hand a very deadly poison is manufactured from the sap of the tree and is used to poison the darts employed in the blow-pipe. As for the tree itself (the botanical name of which is *Antiaris toxicaria,* although there are probably at least two different species in Borneo), it is a Forest Giant rising to the height of a hundred feet or more, and a diameter of four to five feet above the buttresses and roots. It has a fruit about the size of a plum, and the bark, which is grey in colour and about half an inch thick, is used for clothing by the Kayans, who make from it an excellent white cloth.

The actual poison, which is known to Punans as "tajam," is the dried juice of the tree, and is obtained by making incisions in the bark and collecting the emulsive pink sap in cups made from palm leaves. When required for use in the blow-pipe it is heated slowly over a fire until it becomes a thick paste of a dark purple colour, and then worked into a thinner paste on a palette with a wooden spatula. It is very often used in conjunction with other poisons, such as *Tuba* (*Derris*) and *Strychnos;* Dr J. D. Gimlette, in his scholarly book, "Malay Poisons and Charm Cures," suggests that this combination of poisons, perhaps incompatible, is based on the principle that if one does not hit the mark another will.

It is a curious but well authenticated fact that the Punans, who are the principal users of the blow-pipe in Borneo, and who not only collect the Ipoh for their own use, but also for other tribes, are able to handle their poisoned darts with comparative immunity. Perhaps this is a case of homeopathy; or, more likely, the Punans may have acquired immunity by constantly handling the poison.

The usual treatment for cases of dart poisoning

is to draw out the dart carefully, so that the point does not break in the flesh, and then to stab the wound so as to make it bleed profusely; a kind of salted fish-paste known as *blachan* is also used. In both the Malay Peninsula and in Borneo the juice of the common thin-skinned lime is sometimes used.

I have already mentioned the use of *Tuba,* in polluting water for fish; the word itself is used for several poisonous plants used by Malays and others for catching fish, the most important being the large shrub known botanically as *Derris elliptica.* It is curious that while the bark and the fibre of the roots are highly poisonous, the stems are only slightly so, while the leaves possess no toxic properties. The root, according to Dr Gimlette, varies in size from about an inch in diameter to a quarter of an inch or less; when newly dug up it is darkish-brown in colour and tough, but it cuts easily, and has a pleasant "clean" smell something like liquorice-root, and a sweet taste. A white creamy fluid is obtained from the root; this, on drying, turns lemon-yellow. When dry, the root yields a slight cloud of powder on being broken. The plant is frequently cultivated in Borneo, and is found planted in patches in the rice-fields. The chief traders in this commodity are the Chinese, who find a good market for it as an insecticide. To man it is only a partial and cumulative poison; when tasted it increases the flow of saliva, and later induces a feeling of numbness about the tongue and soft palate; ultimately speech is affected. In view of the increasing spread of democracy, it is possible that up to now its value has been underestimated.

Another well-known poison plant found not only in Borneo, but throughout India and Malaya, is the *Datura.* This is a plant belonging to the great Solanum family, and it is therefore a cousin of such well-known plants as the potato, the tomato, " deadly nightshade " (from which Belladonna is obtained), chillies (capsicum), and the aubergine or egg-plant. The Malay name for Datura plants is *Kechubong,* and two varieties are commonly found in Borneo, the Black Datura (*Kechubong itam*), and the White (*K. puteh*). The Black Datura is a quick-growing herbaceous plant about four feet high, with wide-spreading branches and very conspicuous trumpet-shaped flowers, violet in colour with a white lining. It produces round thorny fruits about the size of a walnut, and containing a large number of closely-packed seeds. The white species is slightly taller, and the flower is smaller and more tubular, and its colour is a dead white. The flower is a very striking one, and appears to have much impressed the early Aryan inhabitants of India, for in the Vayu Purana—which dates back to the time of the Guptas (the fourth century before Christ)—it is spoken of as the symbol of the sacred Himalaya mountains; and a conventional Datura flower in stone is constantly found as a gargoyle in South Indian temples.

The poisonous properties of this plant reside chiefly in the seeds, which are very much like those of the common red chilli; no doubt this similarity has often been turned to a practical use, as red chillies, especially in the form of "red pepper," play such a large part in Oriental cooking.

The effect of the poison is usually to produce stupefaction, followed by a form of delirium, in which the victim is half-conscious and possessed by hallucinations. It is therefore a very useful stand-by for robbers and other evil-doers; it was used by the Thugs in India in the early nineteenth century, and it is said that even at the present day there is a sect in Egypt which drugs and then robs country visitors to Cairo.

A well-known fruit which, without being actually poisonous, is used in poisoning is the Papaya or Papaw (*Carica Papaya*), a fruit introduced from South America, but grown commonly by Malays and others in their gardens. The fruit is sometimes known as the "tree-melon." The fruit itself has digestive properties, due to the presence of a drug known as "papain," a ferment similar to pepsin, which is contained in the juice of the tree and of the unripe fruit. The fresh leaves of the tree render meat soft and tender in a wonderfully short space of time—a useful fact to remember where meat has to be eaten almost as soon as it is killed. On the other hand it is a curious fact that on non-meat-eating peoples (such as many of the Hindu castes) it has an irritating and almost corrosive effect; it is also said to cause abortions. In certain parts of Malaya it is used as a poison mixed with the juice of the horse-radish tree (*Moringa*) and the white of a lizard's egg.

In justice to the Papaya, and in spite of Mr C. F. Aflalo's condemnation of it as " an insipid vegetable," it must be said that most people consider it one of the most delicate fruits of the Orient, as superior to the ordinary water-melon as salmon is to cod.

Herbs and drugs for the producing of miscarriages are fairly common; among them a variety of *Plumbago,* known as *Cheraka,* and common in Malay villages. For its illegal purpose the root is mixed with the root of henna, the well-known shrub used in Asia to redden the finger and toe-nails, and in Europe as a dye for the hair; the root of the Champak, whose beautiful scented orange and white flowers are held sacred in India and in Malaya are commonly worn by women in their hair; the root of an evergreen tree called *kenanga,* from which the well-known Malay perfume *ylang-ylang* is obtained; and the root of a flowering tree, *kenerak.*

Among poisons which affect the skin may be mentioned the *Keladi* (*Caladium*), a general name given to a number of lilies of the arum family, one of which, the Cuckoo-pint, is mentioned in Gerard's

Herbal (1597); "it chappeth, blistereth, and maketh the hands rough and rugged, and withal smarting." Another plant which, when it touches the skin, sets up acute dermatitis, leading sometimes to ulceration, is a big tree known as *Rengas,* from which a fine timber known as Bornean Rosewood or Singapore Mahogany is obtained. The wood is of a rich red colour streaked with black; it takes a fine polish, but its value is considerably diminished by its brittleness; and for building purposes it is not popular because of the poisonous black resin which it exudes. According to Dr Gimlette, furniture made of it very often has bad effects when it begins to get old, worn out, and dusty, by inducing an irritation of the mouth, nose, and throat. He mentions also that the eruption produced by the sap is similar to the blisters caused by Mustard Gas.

The root of the Caladium is cooked and eaten as a vegetable by the inland tribes of Borneo. It is also hung on the doorways of rooms to denote that some ceremony is going on in the room, to which outsiders cannot be admitted. The Kayans give it the name "*Long,*" Keladi being the Iban term.

Not only the vegetable kingdom, but also the mineral and animal kingdom provide simple but efficacious poisons, in a manner that reminds one of the witches of "Macbeth." Slugs and snails, grasshoppers and millipedes, frogs and toads, tortoises, snakes and fish are in the pharmacopœia, either by themselves or conjointly with other poisons. The *Ikan keli,* or cat-fish, has a particularly bad reputation and Malays refuse to eat it, considering it unclean. The gall and the slime from its skin are said to be combined with datura and opium for internal application. The pectoral spines of this fish certainly cause septic wounds, but whether the poison is inoculated from the slime or from a definite poison-gland at the base of the spines, has not been determined.

Toads are usually looked upon with disfavour, and it is a scientific fact that toads do secrete two kinds of poison,* one of which brings about a kind of paralysis, while the other affects the heart. Curiously, the flesh of the toad is otherwise not merely safe to eat, but is said, by Dr Gimlette, to

* Butotaline, an acid; and bufotenine; both found by analysis by M. M. Phisalix and Bertrand.

be as nutritious as the flesh of frogs. Personally, I am prepared to leave it at that.

There are a number of hairy caterpillars, the barbed hairs of which cause a painful irritation of the skin; some of these are said to be combined with bamboo filaments for internal use. There are two kinds of these caterpillars recognised by Malays and natives of Borneo; *ulat bulu darat* and *ulat bulu laut,* the hairy caterpillars of the inland and coast respectively. Other insects and reptiles are used more as "excipients", or vehicles of poison, than as in themselves harmful.

Borneo is infested with snakes of various kinds, several of them poisonous; but there are none which deserve special mention. There is, however, one most interesting point that I have observed, namely, that snake-poison may be used, homeopathically, as its own antidote. On one up-country expedition I was surprised, as we were going along, to see one of the men who was carrying a load of rice suddenly drop down and get rid quickly of his load, and chop at something in the path. A poisonous snake had been lying hidden among the leaves, and had bitten him on the foot. It was a green viper, which was well-known to all of us to be very poisonous, and the bite often fatal. I got a bandage out quickly and made a sort of tourniquet round the calf of his leg, whilst his companions were cutting out the gall bladder of the snake and puncturing the wound so as to make it bleed considerably. They then hurriedly rubbed into the wound all the gall which the snake's bladder contained, covering over the surface of the bleeding wound with the skin of the bladder itself.

Contrary to the expectation of his friends the man did not faint, and after stopping with him for about an hour he appeared to be well enough to move about slowly, so we decided to continue our journey as far as our camp, the rest of the party having divided up his load between them, and he himself walking slowly with a stick.

When he arrived at the camp he lay down quietly on his mat and seemed quite contented and happy for the night. The next morning the wound was washed and dressed, and he remained in and around the camp doing odd jobs, such as spreading out wet clothes in the sun to dry, and other light work.

I was at a loss to know whether the tourniquet

had done any good or not, or whether the gall of the snake itself was the real neutraliser of the poison, or perhaps both were helpful.

There are many theories regarding the proper treatment of snake bites, but it has long been recognised that the most important thing to do is to apply a tourniquet and make a free incision immediately over the wound, thereby promoting a primary hæmorrhage, which of course will tend to wash away the injected venom. In this case, however, the cure may have been due to the well-known fact that the body-fluids of a poisonous snake develop antibodies to counteract the poison, thereby causing a natural immunity. Consequently if these fluids were injected into the blood-stream of the victim, there is no doubt that an artificial immunity would develop. This condition, however, takes some time, and there are many factors, such as hæmolysis, sterilisation, etc., to be considered before such a procedure should be attempted, and it is quite possible that the free bleeding combined with the curative power of the gall, which is usually sterile, was the principal factor in the man's complete recovery.

He was apparently quite well on the second day

after it had happened, and the wound quickly healed up.

It should be said, in justice to the pagan peoples of Borneo, that, unlike the Malays of certain countries, they are not naturally given to poisoning or sorcery; in fact a poisoner is considered as something unclean and loathsome, and some years ago there were serious disturbances in the north-east corner of Sarawak owing to a suspicion attached to certain Murut tribes. On the other hand, magic, both black and white, is extensively practised; and, as regards drugs, the *Dayong* or Medicine-man is held in special esteem. Punans and Kalabits have a deserved reputation for their skill as practical, if empiric, physicians; for it is a commonplace that in Borneo as elsewhere among primitive peoples, valuable specifics, unknown to Western medicine, have been in almost daily use for centuries. After all, the medicine-man acts according to his lights and his experience, even if he does not realise the value of careful diagnosis; and as for faith-healing, mesmerism, homeopathy, and (if one wishes) psycho-analysis, the West has a good deal still to learn from the East.

INDEX

INDEX

INDEX

INDEX